W9-CFH-176

OXY

LIBRARY IN A BOOK

STEM CELL RESEARCH

David E. Newton

To the members of the Class of 1951,
Central High School, Grand Rapids, Michigan
Thanks for the memories!
You meant more to me than you ever knew.
And special thanks to Gayle, Fred, Lois, Audrey, and Bonnie
for keeping us all together

STEM CELL RESEARCH

Facts On File, Inc.
An imprint of Infobase Publishing
132 West 31st Street
New York NY 10001

Library of Congress Cataloging-in-Publication Data

Newton, David E.
Stem cell research / David E. Newton.
p. cm.—(Library in a book)
Includes bibliographical references and index.
ISBN-10: 0-8160-6576-4 (hardcover)
1. Stem cells—Research—United States. 2. Stem cells—Research—Government policy—United States. I. Title. II. Series.
QH588. S83N49 2007
616′ .02774—dc22 2005032803

Facts On File books are available at special discounts when purchased in bulk quantities for businesses, associations, institutions, or sales promotions. Please call our Special Sales Department in New York at (212) 967-8800 or (800) 322-8755.

You can find Facts On File on the World Wide Web at http://www.factsonfile.com

Text design by Ron Monteleone
Diagrams by Melissa Ericksen

Printed in the United States of America

MP Hermitage 10 9 8 7 6 5 4 3 2 1

This book is printed on acid-free paper.

CONTENTS

PART I

OVERVIEW OF THE TOPIC

CHAPTER 1

AN INTRODUCTION TO STEM CELL RESEARCH AND ISSUES OVER ITS USE IN THE UNITED STATES

Stem cell research has burst on the national political scene in the United States only within the last decade. Stem cells are special types of cells with the potential for developing into any one of the 200 or more different types of cells found in the human body: skin cells, nerve cells, muscle cells, liver cells, blood cells, and heart cells, for example. This potential raises the possibility that stem cells may be used in bringing about revolutionary changes in medicine, drug testing, and a number of other fields of research.

Although research on stem cells is not new—scientists have been studying them in one form or another for more than two decades—they have become a topic of political, scientific, and general interest only very recently, with the discovery of human embryonic stem cells in 1998. Some people saw these discoveries as heralding a new day in biological and medical research, foreshadowing the development of new materials and procedures that could bring relief to millions of people around the world suffering from diseases that are currently intractable of treatment by any other means. Other people viewed these discoveries with apprehension, recognizing that human embryonic stem cell research can proceed only with the destruction of human embryos, a practice they equated with the murder of human beings.

The debate over stem cell research has been waged on many fronts. Scientists, for example, disagree as to the relative potential value offered by different types of stem cells, those obtained from embryos, from fetuses, and from adult humans. By far the most acrimonious debates, however, have focused on ethical questions. Some people view the fertilized egg and the entity into which it develops over a period of days and weeks as a living human being, alive from the moment of conception. Other people acknowledge the "potential for life" in such entities, but do not accept their designation as fully and completely human. These differences define profound distinctions in the way both groups view research in which these very young entities are used.

The question as to how fertilized eggs, embryos, and fetuses are to be treated by researchers has been resolved in various ways at different times in varying parts of the world. The British Parliament, for example, appointed a committee to study issues surrounding embryonic research in 1985, a committee later to be called the Warnock committee after its chair, Dame Warnock. By 1990, Parliament had adopted legislation defining the conditions under which research on human embryos and fetuses could be conducted and had appointed a national agency, the Human Fertilisation and Embryology Agency, to oversee such research. While many other countries followed the British lead in adopting legislation on embryonic and fetal research over succeeding years, other countries, including the United States, did not. In those countries, policies on human research are often set by a combination of laws on related issues, administrative orders, regulations, and other quasi-legal directives.

Perhaps the most difficult problem about stem cell research is that it raises questions about fundamental beliefs about life and death, about what constitutes a human being. People tend not to be very flexible on such basic aspects of their personal system of ethics and morality, making compromises among competing arguments difficult. And that is the position of the debate over stem cell research in the United States today. The country operates under a general set of regulations set by President George W. Bush in a public address on August 9, 2001, restricting the types of stem cell research that can be funded by the federal government. Yet, much of the research prohibited by the president's policies continues in the United States and other parts of the world, financed here by private funds and overseas by both private and public funds. And individual states, such as California, New Jersey, and Connecticut, have largely rejected the president's arguments in support of his ban on research funding and have adopted legislation that allows state funds to be used for such research.

Thus, stem cell research of all kinds is occurring in the United States and other parts of the world, but under different systems of funding and different types of governmental control. It seems likely that the dispute over stem cell research will continue for the foreseeable future and that diverse approaches to the conduct of such research will also endure.

Stem cells are cells with two essential characteristics: (1) the ability to proliferate—to continue reproducing without changing—for weeks, months, or years, and (2) the tendency to undergo changes during which they differentiate into one or another of the more than 200 different kinds of cells of which the human body is made. The excitement about stem cell research (SCR) arises because of the seemingly endless number of applications it may have in biological research on the fundamental character of cells, in drug research and development, in medical therapeutics, and in other fields. The controversies that have developed in regard to some forms of stem cell research are based on the necessity of destroying entities that are regarded by many people as living organisms—fertilized eggs, embryos, and fetuses—in order to conduct such studies.

What is stem cell research? On what scientific principles are its potential applications based? What scientific challenges face researchers in the field? What ethical, legal, economic, and other questions are raised by stem cell research? These and related questions are the subject of this chapter.

STEM CELLS: THE ALMOST CERTAIN UNCERTAINTY

One of the fundamental problems in discussions of stem cell research is deciding precisely what the term *stem cell* means. According to one writer for the American Bioethics Advisory Committee, "the phrase 'stem cell' or 'stem cells' actually came from histologists [and] . . . has been in the histology texts for a long, long time."[1] Other writers have traced the use of the term to texts dating to the late 19th century, such as E. B. Wilson's classic study, *The Cell in Development and Heredity* (republished by Garland Publishing in 1987), although with a variety of different meanings.[2] Still, other researchers believe that they invented the term, or some variation of it. For example, a publicity release by the Jackson Laboratory in Bar Harbor, Maine, claims that one of its preeminent researchers, Leroy Stevens, was responsible for "the origin of the term 'stem cell'" in 1970, while the writer Ann B. Parson thinks that credit for a modern definition of the term may go to researchers Ernest McCulloch and James Till who spoke and wrote about "stem cells" in their 1960s research on blood-forming cells found in the bone marrow.[3]

Whatever the history of the term, many biologists have long been convinced that cells with the special properties now attributed to stem cells must exist, even if no one had ever seen one. The evidence for the existence of these types of cells falls into three major categories: (1) the ability of many plants and animals to regenerate one or more body parts, (2) the ability, in particular, of the human body to continually regenerate cells, tissues, and organs, such as blood cells, skin, and the lining of the gut, and (3) the existence of teratomas, peculiar tumors that contain tissue and organs from all three embryonic germ layers, endoderm, mesoderm, and ectoderm.

REGENERATION IN LOWER ANIMALS

Many people who have grown up in contact with the natural world have learned that many kinds of animals—salamanders and newts, starfish, segmented worms, zebrafish, hydras and other polyps, and tadpoles among them—are capable of regenerating body parts lost in accidents or, in some cases, removed by curious children or inquisitive scientists. Some of the earliest controlled experiments on regeneration were conducted by French physicist René Antoine Ferchault de Réaumur (1683–1757), who is probably better known for his invention of the alcohol thermometer and a temperature scale named in his honor.

Réaumur's interest in regeneration appears to have started as he walked through the marketplaces of Paris and noticed that many of the crayfish and crabs on sale there had claws of different sizes. He hypothesized that this phenomenon was due to the loss of one claw by the animals, which then regrew a substitute claw that had not yet reached its mate's size when captured for the marketplace. In a paper presented to the French Academy of Sciences in 1712, Réaumur expressed regret that humans had not been granted the ability to regenerate limbs like that possessed by crayfish and crabs, but that they should be glad that nature had given them "a beautiful opportunity to admire her foresight" in providing the lower animals with this ability.[4]

Réaumur was hardly the only researcher to become interested in the process of regeneration. In fact, one modern scholar has observed that "[a]lmost everything that moved in Europe was amputated" by one scientist or another in order to learn more about the process.[5] Réaumur's work, for example, became an inspiration for Swiss scientist Abraham Trembley (1710–84), who took on the daunting task of dissecting a freshwater polyp now known as the hydra. Trembley's careful experiments on the tiny hydra conducted well over two centuries ago have become classics in their field, providing some of the most exhaustive and detailed information about regeneration available for more than a century.[6]

Most researchers were not satisfied simply to experiment on animals and observe regeneration. They also were eager to find an explanation for the phenomenon. By what mechanism did an organism as simple as a hydra "know" how to reconstruct an arm that had been removed? How did a small part of its body "know" how to regenerate all the other parts from which it had been removed? What "knowledge," in particular, did these so-called "lower" animals have that, compared to the far more complex human bodies, to a considerable extent lacked?

These questions intrigued scientists for two primary reasons. First, they hoped that by finding answers to such questions, they might learn to help human bodies regenerate also. Ann Parson speculates that one motivation for Swiss biologist Charles Bonnet's research on regeneration in salamanders—their abilities to regrow an eye, for example—may have been his own failing eyesight. Certainly other researchers of the time who were losing their sight, their hearing, or other mental faculties might have been excused for fantasizing about the possible applications of the process of regeneration not only to the advancement of science, but also to their own well-being, a dream often expressed in today's world by those who see stem cell research as a way of solving medical problems.

Second, regeneration researchers also hoped to answer perhaps the most profound question in all of human biology: How does development occur? What are the processes by which a single fertilized egg grows and develops into all the different cells, tissues, and organism that make up a mature organism?

Most early researchers on regeneration attempted to use their experimental results to support one philosophical view or another about the mechanisms of human development. For example, German biologist Johann Friedrich Blumenbach (1752–1840) attributed to the freshwater polyps on which he experimented a life force that he called *Bildungstrieb*, or "formative drive." Like many scientists of the time, Blumenbach believed that many natural phenomena could be explained only by the presence in plants, animals, minerals, and other materials of a "vital force," which he himself admitted was an "occult quality" that could not be defined in scientific terms but served merely to "designate a peculiar power" to explain the phenomenon's occurrence.[7]

Interestingly enough, modern-day stem cell researchers interested in the process of regeneration find themselves in a situation similar to that of their 19th-century colleagues. They have accumulated large amounts of data about the role of stem cells in the regeneration of cells, tissues, and organs. But they are still largely ignorant of the mechanisms by which that process occurs. The terminology of the problem has changed, surely. Instead of trying to understand "vital forces," researchers are attempting to learn how stem cells "know" how to make new cells and which ones to make. But discovering basic causes eludes modern scientists nearly to the extent that it did Réaumur, Trembley, Blumenbach, Bonnet, and their colleagues.

REGENERATION IN THE HUMAN BODY

The regeneration of human cells and tissues is also a familiar phenomenon to most people. Each time a person cuts a finger, for example, he or she knows that the cut will eventually heal as new skin grows over the wound. A few weeks after the cut, there is usually no evidence that injury ever occurred. Regeneration has occurred spontaneously to replace an essential body part: skin.

The regenerative process of perhaps greatest interest to biologists has long been the production of new blood cells in the body. Blood cells tend to have a very rapid turnover rate in the human body. Some kinds of white blood cells live no more than a few hours (although other types of white blood cells live much longer), while red blood cells die after about 120 days, and platelets survive only seven to 10 days. At these turnover rates, the average human body loses more than a billion blood cells a day. This loss means that the body must have a very active blood-making system, a system that has been found to lie in bone marrow, the spongy tissue that makes up the innermost core of bones.

Long before stem cells were hypothesized or found in any other part of the body, scientists were hypothesizing the existence of something like them in bone marrow, primitive cells that were capable of changing into red and white blood cells and platelets. For example, Russian hematologist Alexander Maximow (1874–1928) carried out a series of precise experiments during the first decade of the 20th century that led him to conclude that there exists a "common stem cell of different blood elements." He thought that this stem cell was the lymphocyte, which, he wrote, "may look different according to the site of residence as well as to the local conditions and which can produce different products of cellular differentiation. The lymphocytes are 'ubiquitär,' of equal value everywhere."[8]

Maximow was wrong about the identity of the blood stem cell, but he was certainly correct in recognizing that some primitive type of blood-making cell existed. The fact that he had picked the wrong stem cell did not make much difference, however, as his work was largely ignored for more than 50 years. In fact, it was not until 1960 that the first real evidence for the existence of a hematopoietic stem cell was obtained. That evidence came from a research team working at the Ontario Cancer Institute in Toronto led by Ernest Armstrong McCulloch and James Edgar Till. McCulloch and Till were studying the effects produced by injecting mice with irradiated and nonirradiated bone marrow. They found no differences in the mice being treated but, incidentally, noted that some mice in both groups tended to develop nodules on their spleens. (Blood formation in mice occurs in both the spleen and in bone marrow.) These nodules were unexpected and, with further analysis, turned out to correspond in size and number with the amount of bone marrow transplanted into the mice. McCulloch and Till eventually concluded that the nodules were produced by the undifferentiated replication of a single "colony forming unit." Before long, they recognized that this "colony forming unit" was, in fact, the hematopoietic stem cell for which hematologists had been looking for so long.

Not only had McCulloch, Till, and their colleagues actually found a stem cell for the first time, but they were also able to provide a definition for the term that has survived to the present day. In a 1963 paper on their discovery, they defined a stem cell as a cell that "should be capable of producing new stem cells like itself; otherwise, it would be on a dead-end street. And it should have the potential to produce different kinds of differentiated cells."[9]

At long last, a real breakthrough in stem cell research had occurred. The first cell to fit that definition—a hematopoietic stem cell—had been discovered. For at least some observers, then, "that, basically, is how stem cell science began."[10]

TERATOMAS

The third line of scientific inquiry that has led to the modern field of stem cell research is the study of teratomas. The term *teratoma* comes from two Greek words meaning "monstrous tumor." Teratomas consist of large clumps of cells typically found in the ovaries or testes, although they may also occur in other parts of the body. They are classified as immature (malignant) and mature (benign) teratomas. The vast majority of cells that make up an immature teratoma are undifferentiated, with a virtually unlimited capacity for growth. Left untreated, an immature teratoma grows so large that it eventually kills its host. This characterization is the classical description of a cancerous tumor, and, for that reason, immature teratomas are also known as teratocarcinomas.

Immature teratomas are quite uncommon, found almost exclusively in humans and especially in adolescent boys and girls between the ages of 10 and 20. One of the most prominent cases of teratocarcinomas in recent years has been that of seven-time winner of the Tour de France bicycle race, American Lance Armstrong, whose condition was diagnosed early enough to permit a successful treatment of his teratocarcinoma.

Mature teratomas, also known as dermoid cysts, are benign tumors consisting of cells that have differentiated into a mixture of cells and tissues, including lipoid (fatty) tissue, neurons (nerve cells), bone, teeth, primitive eyes, hair, and other body structures derived from all three germ lines—ectoderm, endoderm, and mesoderm. Researchers have reported finding virtually all types of cells and tissues in mature teratomas, including partially formed eye sockets, groups of pulsating cardiac cells, and clumps of hair. The bizarre appearance of a mature teratoma under the microscope accounts for its having been named a "monstrous" tumor.

Teratomas have been known and studied for many centuries. The 19th century was a period of particularly intense research, as biologists gradually developed the skills to dissect and analyze teratomas. One of the premier teratologists of the early 19th century, French biologist Isidore Geoffroy Saint-Hilaire (1805–61), clearly understood the biological importance of teratomas or, as they were then known, "monstrosities," beyond their basic appeal as biological peculiarities. Teratomas, he wrote, were a vital key to understanding the process of biological development itself. They are, in some respects, he wrote,

> *permanent embryos; they show us the emergence of simple organs just as in the first days of their formation; as if nature had halted its course in order to provide our too slow observation with the time and means to apprehend it. In the future, therefore, the science of monstrosity cannot be separated from embryogenesis; it will usefully contribute to its progress and will receive no less considerable services in return.*[11]

Unfortunately, the technical skills needed to follow up on this prediction were not available to Saint-Hilaire and his colleagues in the early 19th century. Indeed, it was more than half a century before the technology for the careful study of teratomas became possible and scientists began to develop a more thorough understanding of the composition, genesis, and development of teratomas.

Probably the leading exponent of the experimental study of teratomas at the end of the 19th century was French biologist Camille Dareste (1822–99), sometimes called "the founder of the experimental science of teratogeny."[12] Dareste posed the question as to whether the forms of development normally encountered—normal development, arrested growth, and excessive growth—are the only ones possible for an organism. Or, he asked, are there other ways an embryo might develop, provided the correct environment and stimuli are provided to direct that development? His research objective was to produce embryos with those environments and those stimuli to see how their development would be affected or, as he said, to make teratogeny "a science of all possible bodies" that might have an "unlimited variability" of forms.[13] Saint-Hilaire and Dareste's views of the role that could be placed by the study of teratomas in promoting the understanding of human embryology were remarkably prescient, even though little was to come of their musings for well over a half century.

In fact, the next episode in the scientific study of teratomas relating to the understanding of embryological development did not occur until the 1950s. Then, a series of quite remarkable discoveries began to emanate from the Roscoe B. Jackson Memorial Laboratory in Bar Harbor, Maine. The author of these discoveries was Leroy Stevens, a young researcher who had earned his Ph.D. in embryology at the University of Rochester in 1952 and then taken a job as a junior researcher at the Jackson lab.

Stevens's initial assignment at Jackson was to explore the relationship between cigarette smoking and cancer, using a large population of mice bred at the Jackson laboratory specifically for research purposes. On one occasion, Stevens noticed that one of the mice in his laboratory had developed an enlarged scrotum. Assuming that a tumor might be present, Stevens sacrificed the mouse and dissected the scrotum. He was amazed to find a mature teratoma, one that contained both skeletal and cardiac cells, the latter beating in unison, as they would in a mature heart.

The remarkable point about Stevens's discovery was that it was one of the few times that a teratoma had been found in a nonhuman organism, and the first time it had ever been seen in a male mouse. The fascination of studying teratomas struck Stevens in much the same way that it had many of his 19th-century predecessors, and he was to spend the rest of his working career (37 years) learning more about these intriguing structures.

The first challenge facing Stevens after finding the unexpected teratoma was to determine its source. After all, his primary research project was on locating the possible causes of cancer, one of which might well be a genetic defect in an organism. Stevens decided to look for a genetic defect in the strain of mice with which he was working (the now-famous strain 129) that might have led to the development of the teratoma. He followed this line by looking at male mice in the 19th day of gestation (19th day after fertilization; the day on which a mouse is born), then the 18th day, the 17th day, the 16th day, and so on. With each dissection, he hoped to find the earliest point at which evidence of a teratoma could be found.

Stevens was faced with a daunting task. The teratoma he had found, he discovered, was a rare occurrence among mice, even within strain 129, which appeared to have an unusually high propensity for the development of teratomas. In fact, he eventually discovered that only one in about 1,000 male mice developed such tumors, meaning that he was going to have to be very patient to work his way backward through the developmental history of many mice before uncovering his aberrant cell, provided it existed!

Finally, 12 years after beginning his quest, Stevens was successful. In 1964, he found the earliest example of an aberrant cell leading to the development of a teratoma. He located an abnormal sperm cell growing along the genital ridge of a mouse embryo that eventually developed into a teratoma. Stevens named the aberrant cell a pluripotent embryonic stem cell, that is, a stem cell present in the embryo of the organism with the capability of developing into a wide range of other cells, including the fat, muscle, cardiac, nerve, skeletal, and hair cells and tissues that he had seen in the first teratoma dissected in 1952.

Over the next two decades, Stevens extended his research on teratomas and embryonic stem cells in an effort to learn as much as possible about them. His next step, after identifying pluripotent embryonic stem cells, was to demonstrate that they were the actual culprits in the formation of teratomas. He did so by transplanting tissue taken from the genital ridge of a young fetal mouse into the testis of an adult mouse. When he did so, he made two discoveries. First, he found that the transplanted tissue began to grow and eventually formed a teratoma, confirming the genital ridge as the source of the abnormal stem cells he had found. Second, he found that some of those cells did not differentiate into teeth, hair, muscle cells, and other specialized cells, but simply divided over and over again, remaining in their undifferentiated stem cell form. Stevens summarized these two conclusions in what has become one of the most famous papers in developmental biology: "Some of these cells continued to proliferate indefinitely and served as stem cells of transplantable teratocarcinomas composed of many kinds of tissues. Teratomas originate from a disorganized population of undifferentiated embryonic cells."[14]

SOLVING THE STEM CELL PUZZLE

Anyone familiar with the history of stem cell research must recognize the striking connections between Leroy Stevens's research on teratomas and the work on monstrosities of his 19th-century predecessors. As one historian has said, "ES stem cells, in other words, were discovered as experimental counterparts of the teratocarcinoma—monstrosities, according to the language of nineteenth century biology."[15]

Still, Stevens's research provided no more than an introduction to the study of stem cells. He had become convinced that such cells do exist, showed where they could be found in mice, discovered how they behaved in various kinds of tissues, and successfully established the first stem cell lines, populations of cells that continue to proliferate for very long periods of time without differentiating.

But Stevens's successes only emphasized the research challenges that remained. After all, he (nor anyone else) had never actually seen a stem cell; a great deal more needed to be learned about maintaining stem cell lines; and no one knew very much about the conditions that caused stem cells to differentiate. The search for answers to those and related questions was to drive stem cell research through the 1960s to the present day.

THE PLASTICITY OF STEM CELLS

Over the next four decades, many of these questions (but by no means all of them) were answered by a relentless groups of researchers. Some clues to the amazing plasticity of embryonic stem cells first began to emerge in the

1970s, for example, with research by Beatrice Mintz, at the Institute for Cancer Research in Philadelphia, and Ralph L. Brinster, at the University of Pennsylvania, among others.[16] Both researchers were interested in the fate of embryonic stem cells removed from teratomas and then transplanted into normal blastocysts.

Leroy Stevens had established the model for such experiments by transplanting embryonic stem cells into "unnatural" environments (environments where embryonic stem cells do not normally exist), such as the kidney or spleen of a mouse. He discovered that the transplanted stem cells seemed to go out of control, developing and differentiating rapidly and forming a mature teratoma. He concluded that the transplanted stem cells were reacting abnormally to having been placed into an environment that was somehow "wrong" for them.

Mintz and Brinster were interested in a somewhat different question: What happens when one removes embryonic stem cells from an immature teratoma and places them into a blastocyst, an environment that might be regarded as a "natural" location for embryonic stem cells? Both researchers reached a similar conclusion. In such circumstances, the embryonic stem cells were incorporated into the blastocyst and began to grow, develop, and differentiate naturally, migrating to various parts of the embryo and developing into normal tissues and organs. In everyday terminology, "cancerous" cells (cells taken from a cancerous tumor) had, when transplanted into the proper environment, grown and developed along with and just like healthy ordinary cells.

These experiments confirmed one of the key facts about stem cells. A stem cell, like all cells, consists of DNA molecules that carry instructions as to how that cell is to behave, what kind of cell it is eventually to become. Those instructions are stored in sections of the DNA molecule known as genes. But all of the genes in a DNA molecule are not "turned on" at the same time. In fact, in the first few days of a fertilized egg's existence and in the stem cells that make up an immature teratoma, the only genes that are working are those that tell the cell to reproduce: divide, divide, divide are the only messages the cell receives from the genes in its DNA molecules.

At some point, something happens to the DNA molecules in stem cells. In some cells, the genes responsible for making a fat molecule "wake up" and the cell begins acting like a fat cell. In other cells, the "neuron genes" get turned on and the cell starts to look like a nerve cell. In still other cells, the "skin genes" becomes activated, and the cell turns into a skin cell.

What Brinster, Mintz, and their colleagues showed was that the DNA in stem cells was just waiting for the correct signals to activate genes. Once implanted into a normal embryo, they started receiving the messages they needed as to how they were supposed to differentiate, and they all ended up producing normal cells that migrated to the correct parts of the embryo's growing body doing what they were supposed to do. They were never "bad" cells (even though they came from cancerous tumors). They had just never been given the correct environment or received the correct stimuli.

Isolation of Stem Cells

Even as Brinster and Mintz were completing their research on stem cell plasticity, other researchers were following another line of inquiry, perhaps the most fundamental problem of all: the isolation of an actual embryonic stem cell. Recall that Stevens had not discussed embryonic stem cells, per se. The cells he first found and worked with for most of his life were stem cells taken from immature teratomas, stem cells more precisely called embryonic carcinoma (EC) cells. Their name reflects the fact that EC cells are collected from a cancerous tumor, a teratocarcinoma, not from an actual embryo. Stevens had never been able to carry his research that far back into the embryonic development of mice. And although EC stem cells have all the characteristics of any other embryonic stem cell (the ability to proliferate endlessly and to differentiate at some point), they were not the "Holy Grail" for research for which scientists soon began searching after hearing of Stevens's accomplishments.

The first researchers to achieve that goal were British geneticist Martin Evans and British anatomist Matthew Kaufman, both at the University of Cambridge, in the United Kingdom. In 1981, Evans and Kaufman announced that they had extracted murine (mouse) embryonic stem cells from the inner cell mass of blastocysts extracted from mice and suspended on culture in a petri dish.[17] They then coaxed the stem cells to proliferate and differentiate both in vitro and in vivo. The term *in vitro* (literally "in glass") refers to any process that takes place in an artificial environment, such as a test tube or a petri dish, while the term *in vivo* refers to any process that takes place within a living organism. Only five months later, a similar result was reported by Evans's first postdoctoral student, Gail Martin, then professor of anatomy at the University of California at San Francisco School of Medicine. Martin had extracted stem cells from the inner cell mass of a blastocyst and then cultured them in a medium obtained from a preexisting teratoma culture. Like Evans and Kaufman, she was able to demonstrate the ability of her stem cells to proliferate and to differentiate. In her paper, she used the term *embryonic stem cell* to describe the cells with which she worked, apparently the first time that term was formally used in a scientific paper.[18]

The isolation of embryonic stem cells from mice was a huge step forward in the field of stem cell research. Perhaps most important, it set the stage for the breakthrough for which researchers were really waiting: the isolation of human embryonic stem cells (HSCs). Although that next step was conceptually easy—the process of removing stem cells from a human embryo was technically not all that different from that of removing stem cells from a mouse—the challenges facing researchers were enormous.

In the first place, the technology for isolating stem cells from any type of mammal is staggering. The blastocyst is very small, no more than about 0.1 mm in diameter, considerably smaller than the period at the end of this sentence. Extraordinary skill and special equipment are needed to dissect a blastocyst and remove the stem cells contained within the inner cell mass. Other technical skills are required to transfer and work with the stem cells once they are removed and to maintain the cells in a nondifferentiating, proliferative culture.

Also, working with human embryos presents a host of moral and ethical questions not involved in research on mice, rats, and other animals. As long as a researcher wants to study murine embryos and stem cells, few people are likely to object or withhold research funding. But moving to a study of human embryos and stem cells is a very different matter. Many people believe that human life begins at the moment of conception, the moment a sperm cell penetrates an egg. In that context, the resulting zygote and the blastocyst into which it develops are both living humans on which one should not experiment.

Finally, even if one does not subscribe to that view and accepts research on the blastocyst as morally and ethically legitimate, there remains the problem of finding enough human embryos on which to conduct one's research. Until the mid-1970s, there was virtually no way that a scientist could obtain a human blastocyst short of extracting it from a pregnant woman, which was virtually impossible to do. No matter the technical or ethical restrictions, then, human blastocyst research was nearly impossible, for all practical purposes, because of the limited supply of blastocysts.

An important technological breakthrough in a field that was, at the time, totally unrelated to stem cell research, occurred in the late 1960s and early 1970s that dramatically changed this picture. In 1968, British physiologists Robert Edwards and Barry Bavister successfully fertilized a human egg with human sperm in a petri dish. The experiment provided, for the first time, the basic technology needed for in vitro fertilization (IVF), the combining of egg and sperm outside the human body in order to achieve reproduction of an organism. The successful application of that technology was not reported for another decade when, in 1978, Louise Brown, the world's first "test-tube baby," was born in Great Britain. Before long, interest in IVF among infertile couples began to grow rapidly and by the year 2002, the last year for which data are available, the Centers for Disease Control and Prevention reported that 45,751 children conceived by IVF technology were born in the United States in the preceding year.[19] That trend was common in many parts of the world. In 2005, for example, the Japanese government reported that more than 100,000 babies had been born by means of IVF technology in that nation.[20]

One side effect of the growing interest in the use of IVF among infertile couples has been the availability of "extra" (often known as "spare") fertilized eggs. Typically, an IVF clinic attempts to produce more fertilized eggs than a couple is likely to need to achieve pregnancy. These fertilized eggs can then be stored in case the couple decides to attempt another pregnancy at a later date. Since the number of fertilized eggs is often greater than the number a couple needs and since many couples never choose to attempt a second pregnancy, IVF clinics often have large stocks of fertilized eggs that will never be used. According to one survey, more than 400,000 fertilized eggs are currently in storage in IVF clinics in the United States alone.[21] In many cases, these fertilized eggs are simply stored until they are no longer viable (capable of developing into an embryo upon implantation), and then discarded.

For researchers who had developed the necessary technical skills for working with human blastocysts and who had no moral or ethical reluctance to working with such entities, the spread of IVF technology removed the last hurdle in the search for human embryonic stem cells. It had become only a matter of time before someone would match Evans and Kaufman's and Martin's success with mouse embryonic stem cells. When that breakthrough came in 1998, it was announced not by one, but by two research teams almost simultaneously.

The first report came from a team of researchers at the University of Wisconsin–Madison Regional Primate Research Center under the direction of James A. Thomson.[22] For their studies, the team used early stage embryos obtained by in vitro fertilization and donated to the researchers by the owners of the embryos. The embryos were cultured to the blastocyst stage, and then a total of 14 inner cell masses (ICMs) were isolated from those embryos. Of the 14 ICMs, five distinct embryos were produced, each of which was used to create a new stem cell line. Four of the five stem cell lines were maintained in a proliferative state for five to six months without change, and the fifth had been maintained for more than eight months at the time the report was issued. In their report, Thomson's team describes the criteria and tests used to prove that the stem cell lines were indisputably human embryonic stem cell lines.

Less than a week later, the second report appeared in the *Proceedings of the National Academy of Sciences*, authored by a research team headed by John Gearhart at the Johns Hopkins University School of Medicine, in Baltimore, Maryland.[23] Although the Wisconsin and Johns Hopkins teams used similar research protocols, they began with different materials. While Thomson had used fertilized eggs produced by IVF procedures, Gearhart obtained his stem cells from the immature gonads in fetuses that had been aborted. One difference in these two approaches was that Thomson's research did not fall within the criteria that would have allowed federal funding of his project, while Gearhart's did. In spite of that fact, both researchers chose to carry out their projects entirely with private funding, obtained in both cases from the Geron Corporation, in Menlo Park, California, to whom commercial rights for the discoveries were awarded.

News of the Thomson and Gearhart discoveries met with mixed responses. A number of people were delighted that this important breakthrough in stem cell research had been made. The chances of advancing stem cell therapy to the point where it could be used to treat a number of human diseases seemed, almost overnight, much more promising. Other people were less enthusiastic about Thomson's and Gearhart's achievements. They feared the pressure for manufacturing human embryos to be destroyed for research and medical therapy had also increased substantially. Enough politicians, religious leaders, scientists, and other people—including President George W. Bush and a substantial number of federal and state legislators—held this view to guarantee that, in the United States at least, questions would continue to be raised about and limitations placed on further stem cell research that used human embryos.

Somatic Cell Nuclear Transfer

For that reason, a discovery announced in 2005 by a team of researchers working at South Korea's Hanyang University under the direction of Hwang Woo Suk was all the more startling and thought provoking. The South Korean team reported that they had created a number of stem cell lines carrying the DNA of a patient with a specific disease or injury capable of treatment by stem cell therapy.[24] The announcement was important for two reasons. First, the stem cell lines were created without the use of a human embryo. Instead, researchers used a process known as somatic cell nuclear transfer (SCNT). That process involves the removal of the nucleus from one cell (the host cell) and the insertion of the nucleus that has been removed from a second cell (the donor cell). In the South Korean experiment, the donor cells were skin cells taken from a patient to be treated, and the host cell was an egg cell from which the nucleus had been removed.

The SCNT process may, for reasons that are not clear, stimulate the host cell to begin reproducing asexually in a process known as parthenogenesis. Parthenogenesis occurs naturally among certain types of animals and can be induced experimentally in a wider number of species. It results in the growth and development of the host cell (the egg) in a pattern indistinguishable from that associated with sexual reproduction that involves a sperm and egg. In the South Korean experiment, host cells were allowed to develop to about day six of gestation, at which point embryonic stem cells from the inner cell mass of the blastocysts were extracted.

The second important point about the experiment was that the stem cells harvested by this process were clones—precise genetic matches—of individuals for whom treatment was needed. Should any of these stem cells have been transplanted into one of the patients, there would be no problem of rejection by the patient's immune system because the stem cells would be identical to cells already present in the patient's body. The problem of rejection of stem cells transplanted for medical therapy has been one of the technical questions troubling researchers.

Any hopes that the South Korean experiment would diminish the debate over the use of embryonic stem cells were, however, dashed when comments on the research began to appear. On the one hand, many people saw great promise in the breakthrough that had been made. For example, Robert Schenken, president of the American Society for Reproductive Medicine, said:

> *We applaud Professor Hwang and his colleagues on this stunning scientific advance. Research of this quality serves to motivate and excite us as researchers and as clinicians.*
>
> *With this announcement the fulfillment of the incredible promise that stem cell research will lead to treatments for some of the most disabling illnesses is that much closer.*[25]

Critics of embryonic stem cell research felt differently, however. For example, Richard Land, president of the Southern Baptist Convention's commission on ethics and religious liberty said that "A cloned embryo is a human being. We should not be the kind of society that kills our tiniest human beings in order to seek a treatment for older and bigger human beings."[26] The method of research developed by the South Korean research team has obviously not proved to be a solution to the debate over embryonic stem cell use.

A REVIEW OF STEM CELL SCIENCE

A great deal still needs to be learned about stem cells before scientists can start achieving the hopes and expectations expressed for them. Still, scientists have learned a great deal about stem cells since they became a major focus of research attention only two decades ago.[27]

First, as pointed out earlier, it is known that a stem cell is characterized by two essential properties: (1) its ability to proliferate repeatedly, over many months or years, without differentiating into a specialized cell; and (2) its tendency to begin differentiation when provided with the proper environment and/or stimulus. Stem cells are also characterized by their plasticity, that is, their ability to differentiate into other kinds of cells. A fertilized egg and a blastomere (a cell formed after a fertilized egg has undergone division, but before a blastocyst has formed) are said to be totipotent because they can differentiate to produce any cell in the body. An embryonic stem cell, by contrast, is classified as pluripotent because it can make nearly all kinds of specialized cells. Adult stem cells, like those found in bone marrow, tend to be multipotent because they can make more than one, but not a large number, of specialized cells.

Stem cells are apparently unique among cells in respect to the way in which they can undergo mitosis (cell division). They apparently can following any one of three mitotic paths:

1. The parent cell can divide to produce two daughter stem cells.
2. The parent cell can divide to produce one daughter stem cell and one differentiated cell.
3. The parent cell can divide to produce two differentiated cells.

Stem cells are named and classified according to their origin:

- Embryonic stem (ES) cells are found in the inner cell mass (ICM) of the blastocyst.
- Embryonal carcinoma (EC) cells are obtained from immature teratomas.
- Embryonic germ (EG) cells are derived from primordial germ cells, present in the gonadal ridge of a fetus. EG cells normally develop into mature gametes (eggs and sperm).
- Adult stem cells are undifferentiated cells present in mature tissue and organs, dispersed among the differentiated cells that make up that tissue or organ. Adult stem cells are uncommon, usually present in a concentration of about one stem cell for every 100,000 specialized cells. Evidence suggests that adult stem cells tend to be somewhat less plastic, less capable of differentiating, than are embryonic stem cells or embryonal carcinoma cells. In recent years, scientists have begun to use the term *somatic stem cell* as a preferred synonym for adult stem cell.
- In addition to these, there are stem cells with characteristics similar to those of adult stem cells that are found in blood remaining in the umbilical cord that is usually discarded after the birth of a baby.

The behavior of stem cells is highly dependent on the environment in which they are placed. One major line of stem cell research has been the search for environments that discourage stem cells from differentiating, while encouraging them to proliferate. Experimental environments that meet this criterion are called feeder cells or feeder layers. Historically, the most common feeder layers have consisted of mouse cells (specifically, embryonic fibroblasts, an early form of connective tissue) irradiated to prevent them from growing or developing. In recent years, irradiated human cells have also proved to be successful as the basis for feeder layers. A feeder layer provides a surface to which stem cells can attach, and cells in the feeder layer release nutrients for the stem cells.

A second major line of research has had the opposite goal: to find environments that coax stem cells to differentiate into one or another kind of specialized cell. Some substances that have been found to be effective in directing the differentiation of stem cells are listed in the table found in Appendix A.

Stem cells are also seldom easy to identify by visual examination alone. They tend to look like other kinds of differentiated cells and are, therefore, usually identified by indirect means, such as by chemical markers that exist on their outside surfaces or by chemicals that they secrete. This has important practical significance because researchers often have a problem distinguishing stem cells from specialized cells that occur together within a tissue or an organ, making it difficult to decide what role, if any, the putative stem cell plays in the tissue or organ.

In addition, the plasticity of adult stem cells has become a very important line of research in recent years. Researchers are eager to find out how much potential for use in regenerative medicine and other applications adult stem cells have. From a practical standpoint, most (if not all) of the ethical objections to embryonic stem cell research can be avoided if researchers use only adult stem cells, and not embryonic stem cells. A growing body of evidence suggests that, in contrast to prevailing wisdom, adult stem cells may have some of the plasticity once thought to be a property of embryonic stem cells only.

For example, some researchers now believe that some adult stem cells may have the ability to dedifferentiate. Dedifferentiation is the process by which a unipotent, mature cell somehow reverts to a more primitive multipotent or pluripotent form. Questions as to whether an adult stem cell can actually dedifferentiate; if so, which ones can do so; and how the conversion occurs are currently the subject of intense study.[28] If adult stem cells can be made to dedifferentiate into stemlike cells, researchers will have a whole new source of such cells to use in their studies instead of embryonic stem cells.

Another line of research related to dedifferentiation is transdifferentiation, the process by which an adult stem cell from one kind of tissue differentiates into a cell of a different type of tissue. Some examples of transdifferentiation that have been reported include the conversion of hematopoietic stem cells into three kinds of brain cells, neurons, oligodendrocytes, and astrocytes; into skeletal muscle cells, cardiac muscle cells, and liver cells; and the differentiation of brain stem cells into blood cells and skeletal muscle cells. Although the process of coaxing an adult stem cell to transdifferentiate into a cell of another type appears to be very difficult, it does in theory provide another option to the use of embryonic stem cells for use in regenerative medicine and other applications.

APPLICATIONS OF STEM CELL RESEARCH

The discoveries of murine and human embryonic stem cells by Evans and Kaufman in 1981 and by Thomson and Gearhart, independently in 1998, were greeted with great hope and optimism by scientists for a number of reasons. In the first place, these and related discoveries opened up the possibility that embryologists would have a much better opportunity to study one of the fundamental unsolved problems in their science: How does an organism grow, develop, and differentiate? The question as to the mechanism by which a single fertilized egg evolves into a complex complete organism has long been one of the most fundamental questions in biology. These discoveries provide scientists with the opportunity of examining the very earliest stages of a living entity, the stem cells from which the rest of the organism would eventually grow, and observing the changes that take place in those cells over time. By studying the differentiation of stem cells, biologists now have the opportunity of discovering those factors inherent within an organism itself and those from its surrounding environment that make possible and direct the evolution of a fertilized cell.

A second possible application of stem cell research is its use in drug development. Many steps must be completed before a new chemical compound can be approved as a drug. After the chemical itself has been invented and synthesized, it must be tested both for toxicity and for effectiveness. The testing process often takes many years, beginning with laboratory animals, such as rats and mice, before advancing to human tests. Stem cells can be used more easily and at less expense in place of rats, mice, and other experimental animals for the early stages of drug testing. Compounds found to be toxic to stem cells would then not be advanced to other stages of testing, saving significant amounts of time and money for a pharmaceutical firm.

The applications of stem cell research most often discussed are those in the field of regenerative medicine. The term *regenerative medicine* has been defined by one medical institution as "the regeneration or remodeling of tissue and organs for the purpose of repairing, replacing, maintaining, or enhancing organ function, as well as the engineering and growing of functional tissue substitutes in vitro for implantation in vivo as a biological replacement for damaged or diseased tissues and organs."[29] The use of stem cells in regenerative medicine is based on the fact that many medical problems arise because essential cells, tissues, or organs in the body are damaged or destroyed as the result of disease or injury.

For example, diabetes is a condition that develops when the body is no longer able to metabolize glucose properly. Glucose is a simple sugar, one of the essential fuels used by the body to produce energy. In a healthy body, cells called β-cells are produced in the pancreas. β-cells produce the hormone insulin, which is responsible for the regulation of glucose levels in the body. If β-cells are damaged or destroyed, they are no longer able to produce insulin, or they produce insulin in insufficient quantities. Damage to β-cells may occur at an early age, when the body's immune system attacks and destroys those cells, a condition known as Type 1, or juvenile, diabetes. β-cells may also begin to die off or lose their efficacy later in life, often as a result of a genetic error, a condition known as Type 2 diabetes.

Spinal cord injuries are another example of damage that may be susceptible to treatment with stem cells. Spinal cord injuries occur as the result of automobile accidents or falls in which a person's body is twisted or contorted in such a way that nerve cells are damaged or destroyed and connections among cells are interrupted. Such injuries present three kinds of problems. First, the damaged neurons (nerve cells) must themselves be repaired or replaced. Second, correct connections between neurons must be restored. Third, the protective coating that nerve cells normally have, called myelin sheaths, must also be reconstructed. These coatings act like the insulation on electrical wiring, preventing neural messages from flowing away from the nerve cells.

The hope among stem cell researchers is that embryonic and/or adult stem cells may be used in treating a variety of medical conditions caused by damage to cells, tissues, and organs such as those that occur in diabetes and spinal cord injuries. A general outline as to how such treatments would be conducted is as follows: First, workers would obtain a source of stem cells, such as the blastocyst of a mouse, rat, or human. Then, stem cells would be retrieved from the inner cell mass of the blastocyst. Those embryonic stem cells would then be cultured in vitro, allowing them to proliferate to some stage. At some point, the stem cells would be coaxed into differentiating into some specific type of cell, such as those in the pancreas responsible for the production of β-cells or the cells responsible for the production of new neurons or myelin sheaths. These differentiated cells would then be injected into the "patient," whether it be a rat, mouse, or human with some type of medical problem. Those cells would then, presumably, migrate to the proper location in the host animal and begin carrying out the function for which they are programmed (making β-cells, neurons, or myelin sheath material) or stimulate existing cells in the tissue or organ to initiate the body's own repair of its damaged part.

This brief description ignores the numerous difficult technical problems still to be solved in carrying out such a procedure. For example, scientists currently have very little idea as to how to make an embryonic stem cell begin differentiating into some desired type of mature cell. Also, they are uncertain as to how likely it is that a transplanted cell will find its way to the precise location in the body and begin operating the way a natural cell in that tissue or organ operates. Also, a number of questions remain as to possible undesirable side effects that might result from such a procedure. What is to prevent a transplanted cell, for example, from wandering away from the injection site to a totally inappropriate organ, such as a β-cell to the brain?

The fact that numerous problems remain to be solved does not mean that scientists have no hope for this procedure, however. Indeed, many new areas of science, like stem cell research, pose apparently limitless and, sometimes, unsolvable problems for researchers. Given enough time, however, many if not all of those problems may eventually be solved.

At this point in history, then, the question about stem cell research might be whether there is any evidence that transplantation of stem cells shows any promise in regenerative medicine. The answer to that question is probably yes. In June 2001, the National Institutes of Health (NIH) issued a report, *Stem Cells: Scientific Progress and Future Research Directions*,[30] describing the use of stem cells in regenerative medicine and outlining some of the progress that had been made thus far in the field. Individual chapters discussed progress in the treatment of autoimmune disorders, diabetes, disorders of the nervous system, and cardiac problems. In each case, authors of the report were able to find examples of preliminary research that might eventually lead to the use of stem cells in regenerative problems.

In one experiment, for example, a team of researchers led by Ron McKay at the National Institutes of Health's National Institute of Neurological Disorders and Stroke, in Bethesda, Maryland, attempted to cure diabetes in mice with murine embryonic stem cells. They cultured those cells until they formed embryoid bodies, clumps of cells that develop when stem cells aggregate with each other during the process of cell culturing. They then selected from the embryoid bodies a certain subset of cells that had some of the characteristics of β-cells. They passed those cells through five culturings until they began to show some of the physical characteristics of the portion of the pancreas where β-cells are produced, the islet of Langerhans. When treated with a solution of glucose, those cells secreted insulin, the way a normal pancreas would. When injected into diabetic mice, however, they appeared to have no effects on the symptoms of the disease.[31] Typical of research on any new problem in science, this study was able to demonstrate one small step forward, in spite of its falling short of a complete success.

A similar report was issued by a group of scientists led by Douglas Kerr at Johns Hopkins University in 2001. In this case, researchers studied the effect of injecting barely differentiated embryonic germ cells into rats with spinal cord injury. The injury was induced in the rats by injecting them with the Sinabis virus, which attacks and destroys motor neurons in the spinal cord. Rats that survive the treatment display greater or lesser restrictions on their ability to walk and move around. Kerr's team began with a culture of embryonic germ cells that they allowed to proliferate until it just began to differentiate. At that point, they identified and extracted a subset of cells that showed chemical markers characteristic of nerve cells. They allowed those cells to develop to embryoid bodies and then injected them into the spinal fluid of the injured rats. Three months after the treatment, rats that had been injected with stem cells were moving significantly better than those who had not received the treatment.

The problem with these results was that researchers were not able to identify the reason for the rats' improvement. On the one hand, the injected cells may have behaved as expected, reproduced, and generated enough new neurons to make up for those killed by the Sinabis virus. On the other hand, the cells may have in some way stimulated existing cells remaining in the rats' spines, causing them to take over the job of repair. In either case, the results were the same and were apparently the result of the cell transplantation procedure. Authors of the NIH study concluded their report with the observation that "[i]nvestigators have shown that differentiated cells generated from both adult and embryonic stem cells can repair or replace damaged cells and tissues in animal studies." The problem thus far has been, however, that "there have been very few studies that compare various stem cell lines with each other," making it unclear as to the precise reason that certain results are observed in each experiment. "Predicting the future of stem cell applications," they conclude, "is impossible, particularly given the very early stage of the science of stem cell biology. To date, it is impossible to predict which stem cells—those derived from the embryo, the fetus, or the adult—or which methods for manipulating the cells, will best meet the needs of basic research and clinical applications. The answers clearly lie in conducting more research."[32]

OPPOSITION TO STEM CELL RESEARCH

Support for stem cell research tends to be widespread. Most people seem to understand and accept the proposition that stem cell technology may provide a significant breakthrough in the treatment of diseases for which there are currently no effective cures. In public opinion polls, a majority of respondents support even the most controversial field of stem cell research, that which involves the use of embryonic stem cells.[33] Respondents in those polls also tended to support the use of federal funds to support research on embryonic stem cells and to a relaxation of federal restrictions on such research.

Opposition to stem cell research tends to focus not on the use of adult stem cells, a practice about which there is relatively little controversy, but on the use of embryonic and fetal stem cells, a procedure to which a number of people have very strong objections. These objections are based primarily on the belief that life begins at the moment of conception, the moment at which sperm and egg fuse to form a zygote. That zygote and the blastocyst, embryo, and fetus into which it grows are, according to this belief, as much alive as any person walking on the Earth today. It is entitled to the same respect and legal protection provided to any living human.

The irony of the present debate over stem cell research, then, is that one of the most contentious, complex, and difficult issues facing the modern world arises out of one basic question: When does life begin? If life begins at conception, then society as a whole and researchers in particular must answer the following question: Is it justifiable to destroy one life (the zygote, blastocyst, or embryo) given the possibility that such an act may bring relief to and, possibly, save other lives? On the other hand, if life does not begin at conception, but at some later point, what is that point, and how does that fact affect the conduct of research on embryonic stem cells?

This question—when does life begin?—is so critical that the terminology used in talking about stem cell research is important. Indeed, it seems possible that more words have been written and spoken about terminology in the debate over stem cell research than for any other social issue of the present day.[34] For example, opponents of embryonic stem cell research talk about "the killing of the most defenseless and innocent of human beings" and "destroying a living human embryo," the first statement by Senator Sam Brownback of Kansas and the second, by President George W. Bush.[35] By contrast, proponents of embryonic stem cell research and others who hope to conduct the debate in a different context may use scientific terms, such as zygote, blastocyst, or embryo, or quasi-scientific words, such as entity or preembryo.

The latter term has an especially interesting history. It was apparently first used in 1979 by American embryologist Clifford Grobstein, who apparently meant for the term to designate the entity that develops from a fertilized egg up to the point at which it implants, about 14 days after conception.[36] Some later observers have suggested that Grobstein introduced the term, shortly after the birth of the first baby conceived by in vitro fertilization, in anticipation of the problems that were to come about in the use of such entities in future scientific research. The term has upset and frustrated opponents of embryonic stem cell research ever since, who argue that it represents a way of hiding from the general public the murder of unborn children for scientific research. One critic, for example, has written that:

> *The so-called preembryo is a false stage (period) of human development invented by an amphibian embryologist for political reasons, only. It has no credible scientific justification. Thus, the inclusion of this term into the language of Human Embryology has become a hoax of gigantic proportion. Adolph Hitler said: "The great masses of people . . . will more easily fall victims to a big lie than to a small one."*[37]

And yet, the term *preembryo* has apparently received wide acceptance in the scientific and legal fields. Many of the court decisions discussed in Chapter 2, for example, make use of the term, as do a number of scholarly articles that discuss the legal and ethical issues associated with stem cell research.[38]

The lesson to be learned, then, is that the choice of words that one uses in talking about stem cell research is hardly an esoteric or academic matter. The language used in the debate can as easily be used to inflame passions and attempt to influence opinion as it is to develop ideas and explain positions.

WHEN DOES LIFE BEGIN?

Various cultures and various religions have attempted to answer the question as to when life begins at different times in history. Discovering those answers, however, is often difficult. Societies frequently have laws and customs that provide clues as to the way they think about the origin of life. But those laws and customs tend to deal with related issues, such as abortion and infanticide, rather than the beginning of life itself. Thus, scholars may be forced to make inferences about beginning-of-life beliefs from limited anthropological and ethnographic data.

Many cultures do have specific standards for determining the moment at which one becomes human. In northern Ghana, for example, a child is said to become a human being seven days after birth. In some parts of rural Japan, "humanness" is said to occur when a child utters its first cry after birth. Among some Native American tribes, the tradition was that the moment of becoming human did not occur until a child began to suckle at its mother's breast. And, in a somewhat extreme case, the Ayatals of Taiwan do not grant personhood to a person until he or she is given a name at the age of about two or three years.[39]

As might be expected, the ancient Greeks were very much interested in the question as to when life begins, partly as a matter of philosophical speculation and partly because of legal issues, such as the legitimacy of abortion. The natural philosopher Plato, for example, argued that the human soul does not enter a person's body until birth. In such a case, abortion could not be considered as murder since the unborn organism was not yet truly a human. In contrast, Plato's student Aristotle believed that the beginning of life is an ongoing process that occurs up to the moment of birth. The embryo and fetus pass through various evolutionary stages until the organism is ensouled and becomes a human. Aristotle, with his usual misogynistic bent, dated this moment at about 40 days after conception for males and 90 days for females.

Christian Views

As was the case in much of scientific theory, Aristotle's views on the beginning of life had a strong influence on Western thought over the next 1,500 years. For example, the great Catholic scholar Thomas Aquinas essentially adopted Aristotle's views and taught that ensoulment began after 40 days of gestation for males and 80 days for females.[40] Prior to those times, the unborn organism was not yet human and, under most circumstances, abortion would have been permitted.

Christian doctrine in the early church was also influenced strongly by Judaic traditions, especially as expressed in the Old Testament. Those teachings were not always clear, but appear to have defined "humanness" as the state at which a child is fully "formed," probably at the time of birth. Because of a mistranslation of this portion of the Old Testament, early Christians adopted the notion of the state of "humanness" as occurring when the fetus (rather than the child) became fully formed, and centuries of debate ensued in which an attempt was made to decide the moment at which that event occurs.[41] That debate was reflected in contrasting positions taken by the church over the beginning of life and, hence, the legitimacy of abortion. In 1140, for example, canon lawyer Gratian compiled the first collection of church laws to be accepted by the papacy. In this collection, he stated that abortion was not prohibited until the fetus was "formed." Although he provides no definition for that term, it is generally assumed to mean the point at which the fetus actually looks like a human being.[42]

Confusion about the moment at which life begins continued within the Catholic church for another seven centuries. In 1588, for example, Pope Sixtus V announced that all abortions were prohibited, and anyone who performed or received one would be excommunicated. That ruling lasted only a short time, however, as Sixtus's successor, Pope Gregory IX restored doctrine to its previous status, permitting abortion on "unformed" fetuses without penalty.

Current Catholic doctrine was established in 1854 when Pope Pius IX announced the dogma of the Immaculate Conception. That announcement made necessary the corresponding decision, contrary to teachings then in place, that human life begins at the moment of conception. This view has remained essentially unchanged with the church since that time.

The teachings on which many opponents of stem cell research in the Roman Catholic church rely for their arguments are drawn from a number of encyclicals and other statements issued by Pope John Paul II during his 27-year reign and by various Vatican offices. For example, in his encyclical *Evangelium Vitae*, John Paul used biblical passages to confirm the view that human life begins at the moment of conception. He quotes Jeremiah 1:5 as one piece of evidence: "Before I formed thee in the belly I knew thee; and before thou camest forth out of the womb I sanctified thee." This passage shows, John Paul writes, that:

> *Human life is sacred and inviolable at every moment of existence, including the initial phase which precedes birth. All human beings, from their mothers' womb, belong to God who searches them and knows them, who forms them and knits them together with his own hands, who gazes on them when they are tiny shapeless embryos and already sees in them the adults of tomorrow whose days are numbered and whose vocation is even now written in the "book of life" (cf. Ps 139: 1, 13–16). There too, when they are still in their mothers' womb—as many passages of the Bible bear witness—they are the personal objects of God's loving and fatherly providence.*[43]

The Pope also alludes to Luke's story of Mary's meeting with Elizabeth, at which time Elizabeth's fetus "leap[ed] in her womb" in recognition of the holy virgin. This story can be interpreted, John Paul says, as "indisputable recognition of the value of life from its very beginning."[44] Based on his studies of scripture and church teachings, John Paul concludes in this encyclical that "the use of human embryos or fetuses as an object of experimentation constitutes a crime against their dignity as human beings who have a right to the same respect owed to a child once born, just as to every person."[45]

John Paul II and other officials of the Roman Catholic church, including the current Pope, Benedict XVI, have been among the strongest opponents of embryonic stem cell research since its beginning. Still, some Catholic theologians, philosophers, and ethicists hold views somewhat at variance with those of official church teachings. For example, in a presentation before the National Bioethics Advisory Commission in June 2000, Margaret A. Farley, Gilbert L. Stark Professor of Christian Ethics at the Yale Divinity School, pointed out that:

> *A growing number of Catholic moral theologians . . . do not consider the human embryo in its earliest stages (prior to the development of the primitive streak or to implantation) to constitute an individualized human entity with the settled inherent potential to become a human person. The moral status of the embryo is, therefore (in this view), not that of a person, and its use for certain kinds of research can be justified.*[46]

Catholics who express this view often point to two factors to support their position. The first factor is the long tradition of the church itself, outlined above, in which the time at which life is said to begin has been set at some time other than the moment of conception. Because of this history, these scholars say, reasonable Catholics can disagree as to whether the early embryo is truly human and deserving of the full protection afforded the postnatal child.

The second factor is science's changing understanding of human embryology. As research advances, scientists realize that the process of becoming human is a developmental process that takes place over many weeks and months and does not occur all at once. Hence, the fertilized egg, the preimplantation embryo, the implanted embryo, and the fetus may all have different levels of physical and, presumably, spiritual development. At an early stage of development, the organism may certainly be due respect, without deserving the full respect and legal protection afforded a living human being after his or her birth.

The Roman Catholic scholar Thomas A. Shannon has integrated this notion of a developing human person into his own analysis of the embryonic stem cell research debate. In one article, he draws on the teachings of medieval Scottish philosopher Duns Scotus (ca. 1266–1308) who distinguished between a "common nature" shared by all humans and an "individual nature" that makes each person the distinct individual that he or she is.[47] The early embryo certainly has the "common nature" qualities shared by all human beings, Shannon says, but it only develops its individual qualities that make it a unique human being over time. For this reason, he concludes that "[s]uch a presentation of human nature in the blastomere is preindividual and prepersonal. And because this is human nature and not individualized human nature (the minimal definition of personhood), I argue that cells from this entity could be used in research to obtain stem cells."[48]

Most Catholic scholars reject contrarian views on embryonic stem cell research like those of Farley and Shannon. For example, John J. Conley, S.J., in an essay for the University Faculty for Life web site says that:

> *Those supporting the view outlined above have indeed discovered a certain truth: that the Catholic Church has no definitive position on ensoulment . . . . However, they have distorted this ambiguity and drawn moral conclusions wrongly deduced from this ambiguity. They have operated a strange reading of Church history and have used this truncated reading to legitimize political movements which the Church clearly and rightly condemns.*[49]

The views of other Christian denominations on embryonic stem cell research range widely from one of full support of the Roman Catholic view (and opposition to all embryonic stem cell research) to one of support for such research. For example, in the question and answer section of its web site, Father John Matugiak of the Orthodox Church in America explains that since "Orthodox Christianity accepts the fact that human life begins at conception, the extraction of stem cells from embryos, which involves the willful taking of human life—the embryo is human life and not just a 'clump of cells'—is considered morally and ethically wrong in every instance."[50] Father Matugiak then directs readers to a longer explication of the church's views presented before the National Bioethics Advisory Commission on May 7, 1999, by Father Demetrios Demopulos, of the Holy Trinity Greek Orthodox church, in Fitchburg, Massachusetts.[51]

Positions taken by Protestant denominations differ as widely as do the denominations themselves. At one end of the spectrum, members of the Southern Baptist Church adopted a resolution at their annual convention on June 16, 2001, in opposition to stem cell research involving the use of human embryos. The statement said that, while acknowledging that as evangelical Christians, Southern Baptists "applaud the relief of human suffering and attempts to cure disease," they "object strongly to the notion that pursuing cures for some ever justifies intentionally destroying other human lives to achieve those cures."[52] The Lutheran Church-Missouri Synod has taken a similar stance, adopting a resolution at its 2001 annual convention stating that "[e]mbryos are not, as some claim, 'potential human beings,' nor are they physical entities absent a soul, but are totally and fully human in every way although in the early process of physical development." Therefore, the church declares that "stem cell research involving the destruction of embryos be rejected as sinful and morally objectionable."[53]

At the other end of the spectrum are resolutions such as those adopted by the United Church of Christ and the Episcopal Church. The former body adopted a resolution supporting embryonic stem cell research at its 2001 general synod in October 2001, the first Christian denomination to take such a stand. The church's resolution called for a letter to the president supporting embryonic SCR, recommended lobbying congressional committees responsible for the field of stem cell research, and requested local churches, associations, and conferences to work in support of such legislation.[54] The Episcopal church's action came at its 74th annual convention in 2003, at which a resolution was adopted supporting the "wider availability of embryonic stem cells for medical research" and encouraging the U.S. Congress to "pass legislation that would authorize federal funding for derivation of and medical research on human embryonic stem cells that were generated for IVF and remain after fertilization procedures have been concluded," with a number of provisions added to safeguard the proper use of those stem cells.[55]

Other Religious Views

In their efforts to construct a policy on embryonic stem cell research consistent with as wide an array of religious beliefs as possible, a number of governmental commissions have called on representatives of non-Christian faiths for their views on the subject. For example, the National Bioethics Advisory Commission invited a conservative rabbi, Elliot N. Dorff of the University of Judaism, in Bel Air, California, to testify before one of its sessions in May 1999. Dorff summarized his interpretation of the Jewish position on stem cell research, based on his studies of religious writings and traditional Jewish law. He explained these sources reaffirm a common view in many societies that personhood does not occur until about the 40th day of gestation. Prior to that time, he pointed out, "[g]enetic materials outside the uterus have no legal status in Jewish law, for they are not even a part of the human being until implanted in a woman's womb, and even then, during the first forty days of gestation, their status is 'as if they were simply water' [a quotation taken from the *Babylonian Talmud*]."[56]

Rabbi Dorff's position was largely endorsed by one segment of the American Jewish community in November 2003, when the 67th general assembly of the Union for Reform Judaism adopted a resolution strongly supporting embryonic stem cell research, calling on the U.S. Congress to provide funding for such research and calling on local congregations to develop programs to educate people about the subject of stem cell research.[57]

Interestingly, support for embryonic stem cell research appears to be one of the few issues on which Jews of all major factions—Orthodox, Conservative, and Reform—appear to agree. In May 2005, a special task force of the Orthodox Union and Rabbinical Council of America, a group of very conservative Jewish leaders who had been studying religious issues surrounding stem cell research for many months, sent an action alert to members of its groups encouraging them to contact their legislative representatives to support House Bill 810, the Stem Cell Research Enhancement Act of 2005. The group based its decision on the traditional Jewish perspective that, it said, places a premium on protecting and saving human life. If embryonic stem cell research has the potential to achieve that objective, then, the group's letter said, "[embryonic stem cell research] ought to be pursued."[58]

The Islamic position on stem cell research is largely similar to that expressed by leaders of Judaism. As in the case of Jewish thought, Islamic teachings are based both on holy writings, such as the Quaran (Koran) and the sunnah (the sayings of the Prophet), and on religious law as it has evolved through the ages. According to this body of principles, known as the shari'ah, a distinction is made between potential life and actual life. The point at which that distinction occurs is generally thought to be the 40th day of gestation, similar to the point chosen by Jewish and early Christian writers. Prior to that time, the fertilized egg and the embryo are not really human, and a case can be made for using them, under very strict conditions, for research that may conceivably provide relief from pain and suffering by living humans. As one religious scholar has written, "it is not only allowed but it is obligatory *(fard kifayah)* to pursue this research."[59]

CURRENT ISSUES IN THE DEBATE OVER STEM CELL RESEARCH

While much of the controversy over stem cell research today focuses on a single fundamental question—When does human life begin?—a number of other points of dispute exist. These points include questions such as complicity of researchers and legislators in the death of potential human lives for research purposes, maintaining "respect" for an embryo and other early living entities before birth, pursuing the promise of adult stem cell research, and recognizing and solving technical and other problems related to embryonic stem cell research.

Complicity

In a radio address on August 9, 2001, President George W. Bush explained that he would allow federal funding for stem cell research on certain existing embryonic stem cell lines that were already in existence. By this decision, he argued that the U.S. government would not be responsible for the destruction of human life (early stage embryos) for the purpose of conducting research, but such research could still continue.

A number of theologians and philosophers pointed out what they saw as the flaw in that argument. According to the doctrine of complicity, government officials were guilty of cooperating in the murder of these embryos, they said, even if they did not take part in the actual act of destruction. The doctrine of complicity says that persons or institutions are guilty of a crime, even if they do not actually participate in the crime provided that they aid or abet that crime in some way.

Scholars often list four types of complicity. In the first case, one may actively participate in an immoral act, such as the destruction of an embryo, an entity that some would regard as a living human being. President Bush was certainly innocent of this type of complicity since he directed that the federal government not fund any research in which embryos are destroyed. The second type of complicity is more indirect. It occurs when a person provides support for some kind of immoral act, gives approval to or benefits from the act, or ignores the act. From this perspective, President Bush and researchers would be considered to be complicit with the murder of embryos since they knew about the destruction of the embryos, gave tacit support to the act by making use of the embryos in research, and, in some cases, benefitted financially or in some other way in the use of the embryos. A third type of complicity arises when one knows that an immoral act is about to occur and does nothing to prevent the act. Finally, a fourth type of complicity occurs when someone protects the perpetrator of an immoral act from the legal, moral, or other consequences of the act.

Clearly, the doctrine of complicity casts a very wide net among stem cell researchers. For those who agree with all four aspects of the doctrine, anyone who knows about, takes part in, benefits from, and/or does not act to prevent the destruction of an embryo is as guilty of the murder of the embryo for research purposes as the perpetrator of the act himself or herself. In brief, there is no moral justification for the destruction of any embryo under any circumstances. As one opponent of embryonic stem cell research has written:

> *"Even if NIH [National Institutes of Health] doesn't grant funds to destroy human embryos, it is encouraging those who do by providing a venue for use of the stem cells. Even without the exchange of money, NIH is producing a "market" for those cells. Furthermore, moral complicity moves with the cells. That is to say, when NIH is standing with its arms outreached to receive embryonic stem cells from those who have destroyed embryos to obtain them, the moral guilt passes from one hand to the next. Those who destroyed the embryos are guilty of homicide (there's nothing else to call it), and that guilt passes to those who knowingly use in their research cells obtained at the expense of embryonic life. The NIH guidelines, to put it quite bluntly, do no less than encourage and sanction the destruction of human embryos."*[60]

Those who would encourage the use of embryonic stem cell research see the issue of complicity somewhat differently. In the first place, they may regard the embryo as a nonliving entity, so that the issue of murder or any other type of immoral act is simply not relevant here. Even if they agree that the embryo is alive, they suggest that researchers are guilty of a different type of complicity, something that ethicist John A. Robertson has called beneficial complicity.[61] That term refers to the fact that the benefits achieved by using the fruits of an immoral act may be sufficiently great to excuse the person's complicity. That is, if a researcher can bring relief to many people as a result of using embryonic stem cells, then his or her complicity in the destruction of the embryo from which those stem cells came is of less importance than the progress made. As one observer has written, "At some point, the doctrine of implied complicity in immoral acts must be replaced with reasoned compassion for the living."[62]

Thus far, as with so many other issues in the area of embryonic stem cell research, little progress has been made in resolving the differences among those concerned about the question of complicity in the use of embryonic stem cells.

Respect for the Embryo

Some people involved in the debate as to when human life begins have a somewhat easier position to state and defend: They say that life begins at conception, and that from that point on, whatever you call the evolving organism, it is as much a living human being as anyone who has already been born. It must be treated in precisely the same way as any living human being. People who argue that life begins at some later date, however, are confronted with the more difficult question as to how to treat the living, but not-yet-human, entity that develops from the zygote to the preimplantation embryo to the embryo to the fetus. With what kind of regard should scientists, legislators, and others view this on-the-way-to-becoming human?

The most common answer given to that question is to say that the growing entity does have the potential for life and it does possess certain qualities of humanness, even though it is not an individual human being. For this reason, these people say, the zygote and preimplantation embryo deserve respect. The phrase "the embryo deserves respect" occurs frequently in reports on stem cell research and the writings of both those who support SCR and those who oppose it. For example, the National Bioethics Advisory Commission in its report on *Ethical Issues in Human Stem Cell Research* concluded that "the embryo merits respect as a form of human life, but not the same level of respect accorded persons."[63]

The question, then, becomes what does the phrase really mean in a practical sense? Various scholars have given differing answers to this question.

The most common view seems to be that the preimplantation embryo deserves more respect than an inanimate object or a clump of living cells, but less respect than a fully formed human being. One should not, for example, buy and sell an embryo, the way one might buy or sell a piece of laboratory equipment. But showing respect to an embryo is still more complex. One exhaustive examination of this issue[64] suggests that an important factor in showing respect to an embryo is to ensure that the research in which it is used will really provide sufficient benefit to justify the termination of the embryo's survival. Another factor may be the viability of alternative methods of carrying out the research. Should adult stem cells eventually demonstrate their value as a substitute for embryonic stem cell in the development of medical therapies, then it would show respect to an embryo for it not to be destroyed for research. Also, the possibility of using the embryo for some other purpose, such as being adopted by an infertile couple, may override its use in an experiment, and respecting the embryo in that case would mean choosing adoption over experimentation.

For some scholars, the bottom line in the debate over respecting the embryo becomes, to some extent, a matter of terminology. When one acknowledges that a fully developed human being "deserves respect," the term refers to a whole set of moral privileges due that person. But when one refers to the "respect" one gives an embryo, that term may have a more symbolic meaning, one in which we recognize the solemn value of humanness, but are willing to withhold a number of protections routinely afforded the fully formed human.

Advancing Adult Stem Cell Research

Since the discovery of murine and, more especially, human embryonic stem cells at the end of the 20th century, many scientists and nonscientists have held high hopes for their potential applications in a number of areas, particularly regenerative medicine. Similar hopes for adult stem cells have been somewhat more restrained. The comparatively less enthusiastic claims for adult stem cells have been based on some fundamentally difficult technical problems associated with adult stem cell research. First, as noted above, adult stem cells are relatively rare in tissues and organs, present on average to the extent of about one adult stem cell for every 100,000 differentiated cells. Indeed, some question remains as to whether stem cell counterparts even exist for all 200-plus cells present in the human body.

Second, adult stem cells tend to have physical and chemical characteristics similar to their mature counterparts, making identification and extraction quite difficult. Third, adult stem cells are usually difficult to maintain in an undifferentiated state than are embryonic stem cells. Finally, evidence suggests that adult stem cells may be less plastic then embryonic stem cells, demonstrating multipotency rather than the pluripotency observed in embryonic stem cells. Problems such as these have made many researchers leery of the potential of adult stem cells to achieve the same practical applications expected of embryonic stem cells.

That presumption may be changing. Over the past decade, a number of researchers have been reporting success in identifying and extracting adult stem cells from tissues and organs, maintaining those cells in a proliferative, undifferentiated state for long periods of time, and using them to treat a variety of medical conditions. One of the most exhaustive reviews of this research[65] reports in a number of murine and human systems, including bone marrow, peripheral blood systems, the nervous system, muscles, liver, pancreas, the cornea, salivary glands, skin, synovial membranes of the knee, heart, cartilage, thymus, teeth, adipose tissue, and umbilical cord and amniotic fluid. The reviewer concludes his article on adult stem cells by saying that "[r]esults from both animal studies and early human clinical trials indicate that they have significant capabilities for growth, repair, and regeneration of damaged cells and tissues in the body, akin to a built-in repair kit or maintenance crew that only needs activation and stimulation to accomplish repair of damage. The potential of adult stem cells to impact medicine in this respect is enormous."[66]

Reports of successful adult stem cell transplants in the treatment of murine and human disorders have appeared not only in the scientific literature, but also in the popular media and in testimony before Congress and other legislative bodies. For example, scientists at the New York Medical College in Valhalla announced in September 2003 that they had successfully treated a patient's cardiac disease by transplanting cardiac progenitor cells from rats into the patient's heart. The transplanted cells apparently stimulated stem cells present in the patient's heart muscle, causing regeneration of the damaged cardiac tissue. The team's leader, Piero Anversa, explained that "We have already identified where the stem cells reside and are developing strategies to mobilize them to migrate to the damaged cardiac site. . . . In time we will be developing Phase I clinical trial protocols for submission to the FDA."[67]

A recent study of particular interest on adult stem cells is one reported at the Scripps Research Institute, in La Jolla, California, in 2003 under the direction of Sheng Ding. The Scripps team discovered a synthetic chemical that has the ability to induce dedifferentiation in somatic cells, restoring their multipotency. Ding's team named the chemical reversine because of its ability to reverse the normal process of cell maturation. In announcing the discovery, Ding predicted that "[t]his [type of approach] has the potential to make stem cell research more practical. . . . This will allow you to derive stem-like cells from your own mature cells, avoiding the technical and ethical issues associated with embryonic stem cells."[68]

One of the most intriguing features about adult stem cell research in the past decade has been the extent to which science, politics, and religion have become so intertwined with each other. The reporting of scientific developments that may have an impact on human life without personal or institutional bias is always difficult. Describing the invention of a new kind of nuclear weapon, for example, without revealing one's own biases about nuclear warfare is always a challenge for a writer. And such has often been the case in reports of developments in adult stem cell research.

When such reports appear in journals, magazines, or newspapers, or on web sites sponsored by religious or political groups with strong feelings about stem cell research, they may become more than just reports. They may become part of the arsenal a particular individual or group uses in arguing against or in support of stem cell research, in general, or embryonic stem cell research in particular.

Adult stem cell research has been a topic of particular interest to groups opposing embryonic stem cell research because it provides a realistic technological option (adult stem cell therapy) to a practice that they abhor (embryonic stem cell therapy). In some cases, such groups may actually sponsor research on adult stem cells,[69] but more often, they enthusiastically report on every new breakthrough in adult stem cell research as part of their campaign against embryonic stem cell research, even when such research has yet to be confirmed by outside researchers. For example, one physician who opposes embryonic stem cell research has provided a review of research on adult stem cell research and concludes that "whoever would cure, must use adult stem cells," and has described some scientists' cautions about the problems to be solved in pursuing embryonic stem cell research as "restrained language used by established science to describe a truly disastrous set of results."[70]

The lesson to be learned from these stories is that scientific developments in stem cell research are sometimes presented in the popular media with a twist that reflects an author's or sponsor's own particular prejudice about the morality of embryonic stem cell research. That statement applies equally to supporters of and opponents to embryonic stem cell research. Neither side is likely to miss an opportunity to point out how results of scientific research that, in and of itself are ethically neutral, confirm its own position on this controversial issue.

The power that political and religious organizations may have in influencing the direction of scientific research is illustrated by the stand of one major health advocacy organization on stem cell research. The American

Heart Association (AHA) is listed in Chapter 8 of this book as an organization that supports stem cell research. But the organization itself funds only adult stem cell research, not embryonic stem cell research. This decision, according to the AHA web page on stem cell research, was based not on the relative merits of both types of research or the science and technology involved, but on concerns about "the impact of funding embryonic stem cell research on volunteer and staff retention . . . [and] the potential adverse financial impact which would restrict the ability to fund other lifesaving research and programs." Without question, every individual and every organization has the absolute right to choose which forms of research to support and which to oppose. Anyone interested in learning more about the subject of stem cell research should be alert for the difference between objective reports of ethically neutral scientific developments and the "spins" placed on those reports by groups with vested interests in one side or the other of the stem cell research controversy.

Technical Concerns with Embryonic Stem Cell Research

Research involving embryonic stem cells, like that with adult stem cells, has experienced some exciting breakthroughs in the past decade. Some of the most important of these discoveries have been in the area of technique and methodology. In March 2005, for example, researchers at the University of California-San Diego (UCSD) School of Medicine reported that they had developed a method for maintaining embryonic stem cells in a proliferative state without the use of feeder cells.[71] The UCSD scientists added a human protein called activin A to an embryonic stem cell line culture and found that the culture survived and divided as efficiently as it had using more traditional (usually murine) feeder layers. This result is important because the use of either nonhuman or human feeder cells introduces the possibility that proliferating embryonic stem cells may become contaminated by those feeder cells, making them useless for transplantation into a human subject.

Only months earlier, a second group of UCSD researchers had described another methodological development in the use of embryonic stem cells.[72] They reported initial progress in coaxing such cells to differentiate along one or more lines of specialization into liver, adipose, nerve, skin, or some other type of mature cell. The news was important because the use of embryonic stem cells in regenerative medicine and other applications depends on just this ability of directing the differentiation of cells in some specific direction.

Researchers have also reported success in using embryonic stem cells to treat certain medical conditions and disorders, usually in experimental animals. In January 2005, for instance, a research team at the University of Wisconsin–Madison's Stem Cell Research Program headed by Su-Chun Zhang reported that they had coaxed human embryonic stem cells to become spinal motor neurons, nerve cells that control movement.[73] This accomplishment is one of the first steps needed for using embryonic stem cells in transplantations to treat diseases of the nervous system, such as amyotrophic lateral sclerosis (ALS, or Lou Gehrig's disease).

In a study that represents a potential second step in such treatments, researchers led by Hans Keirstead at the University of California–Irvine's Reeve-Irvine Research Center reported in May 2005 that they had treated rats with spinal cord injuries with human embryonic stem cells and observed significant improvement in the rats' motor skills.[74] Rats who had been injured up to seven days before treatment experienced nearly complete recovery after treatment with ES cells, although treatment was not effective with animals who had been injured many months prior to treatment. The principal investigator later said that "We're very excited with these results. They underscore the great potential that stem cells have for treating human disease and injury."[75]

That level of enthusiasm is not uncommon among scientists working with embryonic stem cells. However, most researchers acknowledge that the bright hopes for using ES cells in regenerative medicine and other applications depends on solving a number of difficult technological problems. One of the most serious of those problems is the possibility that embryonic stem cells implanted into a patient may begin to proliferate and grow out of control, as happens during the development of an immature teratoma. Of course, scientists already know that such events have occurred in experimental animals following the implantation of embryonic stem cells, and the question remains as to how it can be prevented in humans during a therapeutic application. Uncontrolled growth of cells would result in the development of a cancer, which could be as or more serious a medical problem as the one being cured.

A second major problem relating to the use of embryonic stem cells is the possibility of rejection by a patient's body. Any time cells, tissue, or organs are transplanted from one person to another person, the recipient's body is likely to initiate an immune reaction against the implanted material. Cells in the transplanted material carry chemical markers on their surface that provide them with a unique identity. A recipient's immune system is able to detect those markers and recognize that they do not come from its own body. It then initiates an autoimmune response that can cause serious damage to, and even kill, the patient. Again, the cure in such a case is worse than the disease being treated.

A third problem relates to maintaining control over the destination of stem cells implanted into a patient's body. Suppose that scientists are able to find a way of coaxing embryonic stem cells into a particular type of specialized cell that is then injected into a patient to treat a disease. Can a scientist be certain that, once injected, those cells then travel to the appropriate place in the body (nerve cells to the nervous system; muscle cells to muscle tissue; skin cells to skin; and so on), or is it possible that they would migrate to inappropriate locations (nerve cells to muscle tissue or skin cells to the nervous system, for example)? Thus far, scientists have been unable to answer that question.

Technical problems like these may exist for adult stem cells also, but are likely to be less serious than they are for embryonic stem cells. The reason is that the very quality that makes embryonic stem cells desirable for so many medical applications—their pluripotency—also contributes to some of the problems that may be associated with their use in medical situations.

LEGISLATION AND PUBLIC POLICY

Disagreements over ethical issues—what *ought* individuals and societies do with regard to certain practices—often evolve into political and legal disputes—what should individuals and societies *be allowed* to do. Such has been the case in the debate over stem cell research. Governments around the world are now trying to decide how, if at all, they should attempt to monitor and control research on stem cells.

GREAT BRITAIN

Probably the earliest attempt to deal with bioethical issues that would later arise in the stem cell debate took place in Great Britain in the early 1980s. Questions as to the status of the human embryo and how it should be treated in scientific research came about shortly after and as a result of the birth of the first baby born as a result of in vitro fertilization, Louise Brown, in 1978. Parliament appointed a committee in July 1982 "to examine the social, ethical, and legal implication of recent, and potential developments in the field of human assisted reproduction."[76] The committee was chaired by Dame Warnock and is, therefore, generally known as the Warnock Committee, and its final report as the Warnock Report.

The major recommendation of the committee's 103-page report, issued in July 1984, was that research should be permitted on human embryos during a period up to 14 days after fertilization, provided a number of conditions were observed. The committee also recommended the establishment of a supervisory body to monitor all such research in the country. Parliament was somewhat slow in acting on the Warnock Committee's recommendation, but eventually passed the Human Fertilisation and Embryology Act in 1990, essentially adopting the committee's major recommendations. The act also created the Human Fertilisation and Embryology Authority (HFEA) to regulate in vitro fertilization and experimentation on human embryos. HFEA's charter restricted it to the approval of activities that fall into one of six categories:

1. promoting advances in the treatment of infertility;
2. increasing knowledge about the causes of congenital disease;
3. increasing knowledge about the causes of miscarriages;
4. developing more effective techniques for contraception; and
5. developing methods for detecting the presence of gene or chromosome abnormalities in embryos before implantation; or
6. for such other purposes as may be specified in regulations.

The Human Fertilisation and Embryology Act had been in force for only a relatively short period of time before scientific developments began to raise new issues regarding the cloning of human embryos. One of the most significant of those developments was the cloning of a sheep, named Dolly, in 1996, by scientists at the Roslin Institute in Scotland under the direction of Ian Wilmut. Dolly was conceived by somatic cell nuclear transfer (SCNT), a form of reproduction not envisioned by the Warnock committee, not mentioned in its report, and hence not specifically covered by the Human Fertilisation and Embryology Act.

The birth of Dolly prompted the British government to take two additional actions to handle new issues raised by SCNT. First, the HFEA and the government's Human Genetics Advisory Commission (HGAC) undertook a review of the government's position on human cloning, requesting input from the scientific community and the general public. As a result of this review, the HFEA and HGAC recommended, among other things, that the Secretary of State for Health should add two new fields of research for which the HFEA might issue licenses: (1) the development of treatments for mitochondrial disease and (2) the development of treatments for diseased or damaged tissues or organs.

The government also created a new advisory committee in September 1999 under the chairmanship of Sir Liam Donaldson, the nation's Chief Medical Officer. The committee was charged with reviewing developments in the field of in vitro fertilization and other kinds of assisted reproduction, determining potential benefits and risks of these developments, and recommending changes in the Human Fertilisation and Embryology Act of 1990 to meet these new findings. The report of the Donaldson committee, issued in June 2000, is important, among other reasons, because of its specific and detailed consideration of the growing significance of stem cell research, a topic that earlier committees had largely not had to deal with.

The Donaldson report agreed in principle with the government's existing position on embryonic research, as expressed in the Human Fertilisation and Embryology Act of 1990 and the policies adopted by the HFEA. It did make nine recommendations for changes in these policies, most notably the addition of three new fields in which embryonic research should be permitted and a firm restatement of the prohibition of cloning for the purposes of human reproduction.[77] In response to the Donaldson report, the British Parliament enacted the Human Fertilisation and Embryology (Research Purposes) Regulations (HFER) of 2001 that added three categories of research for which the HFEA was responsible: (1) increasing knowledge about the development of embryos, (2) increasing knowledge about serious disease, and (3) enabling any such knowledge to be applied in developing treatments for serious disease. The new regulations went into effect on January 31, 2001.

The deliberations within Parliament on the status of the human embryo were, of course, of considerable interest to a number of groups whose position was that human life begins at conception. For such groups, the Human Fertilisation and Embryology Act and related legislation and regulations were incorrect because they authorized the destruction of human life. Passage of the HFER of 2001 prompted one of those groups, the ProLife Alliance, to ask for a judicial review of the government's position on embryonic research. On November 15, 2001, the British High Court ruled that embryos created by SCNT were not regulated by the new HFER because they were not created by fertilization of an egg by a sperm.

The court's ruling raised a host of new questions about the government's system of regulating human embryo research. It meant that a possibility existed that the cloning of humans for reproductive purposes using SCNT might be permitted and beyond the reach of existing regulations. As a spokesperson for ProLife said, "The Human Fertilisation and Embryology Act 1990 is now in tatters . . . the Human Fertilisation and Embryology Authority, which has become the mouthpiece of vested biotechnology interests, has been seriously undermined, and the Government's gross incompetence has been exposed."[78]

The government responded immediately. It announced that legislation designed specifically to prohibit human cloning would be introduced and that the High Court's decision would be appealed to the Appeals Court. The promised legislation was rushed through Parliament, introduced on November 21, 2001, and passed into law on December 4, 2001. Only a month later, the Appeals Court overturned the High Court's decision and restored the original intent of the Human Fertilisation and Embryology acts.

Still, the story was not complete. At this point, ProLife appealed its case to the House of Lords, asking that it rule against existing policies. In response, the House of Lords appointed yet another committee to investigate the social, ethical, and legal implications of embryonic research. That committee issued its report on February 13, 2002, once more essentially upholding government policies and practices on cloning, in vitro fertilization, and embryonic stem cell research developed over the previous two decades. Among its 27 recommendations, perhaps the most important were:

- "To ensure maximum medical benefit it is necessary to keep both routes to therapy [using adult and embryonic stem cells] open at present since neither alone is likely to meet all therapeutic needs."
- "For the full therapeutic potential of stem cells, both adult and ES, to be realised, fundamental research on ES cells is necessary, particularly to understand the processes of cell differentiation and dedifferentiation."
- "Whilst respecting the deeply held views of those who regard any research involving the destruction of a human embryo as wrong and having weighed the ethical arguments carefully, the Committee is not persuaded, especially in the context of the current law and social attitudes, that all research on early human embryos should be prohibited..."
- "Fourteen days should remain the limit for research on early embryos..."
- "Embryos should not be created specifically for research purposes unless there is a demonstrable and exceptional need which cannot be met by the use of surplus embryos..."
- "Although there is a clear distinction between an IVF embryo and an embryo produced by CNR [SCNT] (or other methods) in their method of production, the Committee does not see any ethical difference in their use for research purposes up to the 14 days limit..."
- "The Committee unreservedly endorses the legislative prohibition on reproductive cloning now contained in the Human Reproductive Cloning Act 2001..."[79]

As is apparent from the preceding summary, British policies regarding in vitro fertilization, cloning, and stem cell research have remained consistent over the past two decades. They encourage and support, both financially and in terms of policy, the use of human embryos developed by a variety of methods in research up to about the 14th day following gestation, and they strongly and unequivocally oppose the cloning of embryos for the purpose of human reproduction.

THE EUROPEAN UNION AND OTHER NATIONS

Other nations of the world have taken positions on stem cell research ranging across the spectrum of permissiveness, from being highly supportive to strongly opposed to such research. In Europe, Belgium's laws tend to mirror those in the United Kingdom, while Denmark, Finland, Greece, the Netherlands, and Sweden all tend to have fairly permissive policies that allow researchers to use spare embryos produced originally for in vitro fertilization. By contrast, Austria, France, Germany, Ireland, and Spain tend to have more restrictive laws and regulations, prohibiting the use of human embryos from in vitro fertilization and other sources.[80] A few countries, including Czech Republic, Luxembourg, Malta, and Portugal, had no specific legislation dealing with stem cell research as of early 2006.

Most non-European nations currently have no laws or regulations about, or tend to take a permissive stand on, stem cell research. China, Japan, Singapore, South Korea, and Taiwan, for example, all have relatively permissive attitudes toward stem cell research, allowing experimentation for therapeutic purposes, but not for the purpose of human reproduction. In all of these countries, government has taken an active role in funding and encouraging such research. One observer from the United Kingdom described his impressions of visits to stem cell research laboratories in a number of Asian countries in 2005: "They are at, or approaching, the forefront of international stem cell research," he said. "During our 14-day visit to China, Singapore and South Korea, we encountered some of the best equipped laboratories, most industrious research teams, and most adventurous clinical programmes that any of us had ever experienced."[82]

The status of stem cell research throughout the world is, to a certain extent, still in a state of flux. South Africa, for example, banned all forms of stem cell research until 2003. Its reason for this prohibition was not particularly concern about the destruction of early human life, but the fear that Western corporations would exploit poor South African women, offering them financial incentives to provide eggs and embryos for their research. The nation's health bill for 2003 laid out a new policy for stem cell research, however. The government decided that the therapeutic advantages of stem cell research outweighed any risk to South African women and agreed to permit stem cell research from embryos up to the age of 14 days. The ban on cloning for human reproductive purposes was, however, reiterated.

United States

No comprehensive legislation similar to the United Kingdom's Fertilisation and Embryology Act of 1990 has ever been passed in the United States. The subject of research on fertilized eggs, embryos, stem cells, and related materials has, however, been the subject of considerable discussion for more than three decades. One of the key events motivating that debate was the Supreme Court's 1973 decision on abortion, *Roe v. Wade*. That decision was important not only because it established policy in the United States on the termination of pregnancies, but also because it raised a number of fundamental questions about the beginning of human life and the rights, if any, possessed by prenatal organisms.

At about the same time that the Court was considering the issues involved in *Roe v. Wade*, a potentially revolutionary new scientific technique known as in vitro fertilization was being developed by scientists in Great Britain and other parts of the world. The potential for this new technology to radically change the way humans think about human reproduction was obvious, not only to scientists, but also to legislators, politicians, and the general public, all of whom recognized that new laws, regulations, and policies might be necessary to handle the new reproductive choices that humans might soon have.

A third factor feeding the debate over human experimentation in the 1970s was the disclosure of an experiment being conducted by the U.S. Public Health Service in Tuskegee, Alabama, on the effects of treatments for syphilis. In this study, a number of subjects had been left untreated while health problems attributed to syphilis grew steadily worse, even though effective treatments for the disease were available. Outrage at the lack of ethical consideration given to subjects led to calls for federal legislation to deal with this problem.[82]

The confluence of these three forces led in 1974 to passage of the National Research Act (P.L. 93-348) which included, among other provisions, for the creation of a National Commission for the Protection of Human Subjects of Biomedical and Behavioral Research. One of the charges to the commission was the determination of basic ethical principles that should underlie the conduct of biomedical and behavioral research involving human subjects. That commission met for four days in February 1976 at the Smithsonian Institution's Belmont Conference Center and then continued its research and analysis over a period of nearly four years. It issued its final report on April 18, 1979, a report that has since been called the Belmont Report.

The Belmont Report laid out a relatively simple set of guidelines to be followed in research involving human subjects. It identified three basic ethical principles: respect for persons, beneficence (that is, protection of a person's well-being), and justice (that is, a guarantee that a person is not treated unfairly because of his or her race, gender, class, or other personal characteristic). The report also proposed three fundamental guidelines that should direct human experimentation: informed consent, comprehension, and voluntariness. In connection with its studies, the Belmont committee also issued a number of documents on related topics, including reports on research on the fetus (1975), research involving prisoners (1976), research involving children (1977), psychosurgery (1977), research involving those institutionalized as mentally infirm (1978), ethical guidelines for the delivery of health services by the Department of Health, Education and Welfare (1978), and implications of advances in biomedical and behavioral research (1978).

Upon completion of its final report, the Belmont committee passed out of existence and was replaced by a new congressionally mandated committee, the President's Commission for the Study of Ethical Problems in Medicine and Biomedical and Behavioral Research. Like its predecessor, the President's Committee devoted more than three years studying a variety of bioethical issues, issuing reports on topics such as defining death (1981), protecting human subjects (1981), whistle-blowing in biomedical research (1981), social and ethical issues of genetic engineering with human beings (1982), making health care decisions (1982), deciding to forego life-sustaining treatment (1983), implementing human research regulations (1983), screening and counseling for genetic conditions (1983), and securing access to health care (1983).

One of the recommendations of the Belmont committee was that the secretary of Health, Education and Welfare appoint an Ethics Advisory Board, whose responsibility it would be to review research proposals involving fetuses and children. Pursuant to that recommendation, Secretary Joseph Califano appointed such a board, called the Ethics Advisory Board (EAB), in 1977. One of the first issues with which the new EAB had to deal with was a research proposal for a study of in vitro fertilization procedures. The board agreed to consider that proposal at its May 1978 meeting, and later approved the research. At virtually the same time the board was considering this proposal, the birth of baby Louise Brown, the first child conceived by in vitro fertilization, was announced in England, and secretary Califano asked the EAB to undertake a more comprehensive review of the social, legal, and ethical issues surrounding IVF.

The board issued its report on this question, "HEW Support of Research Involving Human In Vitro Fertilization and Embryo Transfer," on May 4, 1979.[83] It reached conclusions similar to those adopted by various British commissions and the British Parliament, namely that research was acceptable from an ethical standpoint up to the 14th day of gestation, provided that a number of conditions were met. Those conditions included informed consent on the part of the embryo donors and evidence that the goals of the research could not be met by methods other than those that required use of an embryo.

The policy outlined by the EAB report and the recommendations it contained were never implemented. Within months of the report's having been issued, a new president (Ronald Reagan) was elected, and a Republican administration opposed to embryo research replaced a Democratic administration that had been generally supportive of such research. Although the Reagan administration did not enact any new legislation on embryo research, by its refusal to continue support of the EAB and its recommendations, a de facto prohibition on embryo research was established.

This restriction on embryo research remained in effect over the next 12 years, through the administrations of presidents Reagan and George H. W. Bush. Throughout this period, a number of scientists and politicians continued to push for an implementation of the EAB recommendations and a restoration of the board itself. In early 1988, for example, James Wyngaarden, director of the National Institutes of Health, appointed a committee to consider the ethical issues involved in research on the use of fetal tissue transplants for the treatment of certain diseases. The committee issued its report in December 1988 and, by a vote of 19-2, recommended that the federal government renew its funding of research on fetal tissue transplantation. The Bush administration decided not to implement that recommendation, and the de facto ban on research involving embryos and fetuses continued.

During the 12-year moratorium, opposition to the ban on fetal tissue research also continued to grow in Congress. By the last year of Bush's term, in 1992, majorities in both the House and the Senate supported an overturn of the ban and legislation to that effect was passed in both bodies. Bush vetoed the legislation, however, and his veto was upheld in the House of Representatives.

Only with the election of a new president, Bill Clinton, in 1992 was there once more a reversal of fortunes for embryonic and fetal research. One of Clinton's first official acts was to issue a presidential memorandum on January 22, 1993, revoking the existing de facto ban on fetal and embryonic research. Simultaneously, Democratic leaders of the Senate and House of Representatives introduced legislation (S. 1 and H.R. 4) designed to codify the president's action and to authorize the expenditure of federal funds for such research. These bills were eventually passed as the National Institutes of Health Revitalization Act of 1993 and signed by the president on June 10, 1993, after which they became codified as Public Law 103-43.

The treatment of fetal and embryonic research were handled in a somewhat indirect manner in the act. Rather than discussing the ethical issues of such research, S.1 and H.R. 4 simply eliminated provisions for the EAB, essentially removing any requirements for federal review of fetal and embryonic research. In response to this change, Harold Varmus, director of the NIH, appointed a new committee, the Human Embryo Research Panel (HERP), charged with developing standards for determining which fetal and embryonic research projects could ethically be supported with federal funds and which could not. HERP issued its report on September 27, 1994, listing areas of research that it regarded as acceptable and unacceptable. Among the approved activities was the creation of human embryos specifically for the purpose of experimentation. Two months later, on December 1 and 2, the Advisory Committee to the Director for NIH met to consider HERP's recommendations. The committee approved all of those recommendations and forwarded them to Varmus.

Having heard about the NIH and HERP proposed actions, however, President Clinton intervened in the process on December 2 and issued an order prohibiting approval of the made-for-research embryo section of the committee's report. "I do not believe," he said, "that federal funds should be used to support the creation of human embryos for research purposes, and I have directed that NIH not allocate any resources for such research."[84] Except for the provision to which Clinton objected, Varmus accepted the HERP report and began the process of implementing its remaining provisions. He announced that, among other activities, the federal government would begin funding research involving so-called surplus blastocysts produced for in vitro fertilization.

Varmus's announcement soon became moot. In 1995, Representative Jay Dickey (R-Ark.) proposed a rider (an amendment) to the appropriation bill for the Department of Health and Human Services. The rider prohibited the use of federal funds for "(1) the creation of a human embryo or embryos for research purposes; or (2) research in which a human embryo or embryos are destroyed, discarded, or knowingly subjected to risk of injury or death greater than that allowed for research on fetuses in utero . . ."[85] That rider has been passed every year since 1995 as part of the appropriations act for the Department of Health and Human Services, under which the NIH is funded.

Current law and policy in the United States, unlike that in most other nations, has evolved not through the adoption of legislation developed through debates and discussions among scientists, politicians, and ordinary citizens within and outside federal legislative bodies, but as secondary actions (such as presidential memoranda and riders attached to appropriations bills) on an ad hoc basis in response to immediate problems.

All of which is not to say that efforts have not been made along those lines, efforts to take the pulse of all interested observers in the stem cell debate in an effort to understand the issues involved in stem cell research and the kinds of legislative and other actions that might be necessary to develop a coherent national policy. For example, on October 3, 1995, President Clinton issued Executive Order 12975, creating the National Bioethics Advisory Commission with two primary functions: (1) to "provide advice and make recommendations to the National Science and Technology Council and to other appropriate government entities regarding the . . . appropriateness of departmental, agency, or other governmental programs, policies, assignments, missions, guidelines, and regulations as they relate to bioethical issues arising from research on human biology and behavior"; and (2) to "identify broad principles to govern the ethical conduct of research, citing specific projects only as illustrations for such principles."[86] During its five-year existence, the commission issued a number of reports on bioethical issues, including the cloning of human beings, research involving persons with mental disorders, ethical issues and policy principles involving human biological materials, ethical issues in human stem cell research, and ethical and policy issues in research involving human participants. The report on stem cell research has long been one of the most comprehensive and thorough discussions of social, legal, ethical, and religious issues surrounding stem cell research yet produced. When the commission's charter expired in 2001, it was not renewed by the incoming Bush administration.

The nearly simultaneous announcements in November 1998 by researchers at the University of Wisconsin and the Johns Hopkins University of the discovery of human embryonic stem cells raised an important new issue for federal regulators. The revolutionary promise of these discoveries was obvious to everyone concerned with stem cell research, but a question remained as to whether the federal government was allowed by existing legislation and regulations to fund research involving human embryos. That question was put to the general counsel of the Department of Health and Human Services, Harriet Rabb, shortly after the Wisconsin and Johns Hopkins announcements were made. On January 15, 1999, Rabb issued her opinion. She concluded that research on human embryonic stem cells was eligible for federal funding because the cells themselves "are not a human embryo within the statutory definition."[87] That opinion was later criticized by opponents of human embryonic stem cell research as overly legalistic and relying on the precise letter of existing law and regulations rather than recognizing the spirit of those rules. In any case, these objections were moot because the issuance of federal regulations is a long process, and regulations for the funding of the controversial research were not issued until August 23, 2000. Only two months later, the nation had a new president, George W. Bush, with a strong objection to the funding of human embryonic stem cell research.

Bush was, of course, aware of the promise and the controversies surrounding stem cell research. Early in his administration, therefore, he established his own bioethics committee, modeled to some extent on the Clinton commission that had just been dissolved. The President's Council on Bioethics was created on November 28, 2001, by Executive Order 13237. Its charge was fivefold:

1. "To undertake fundamental inquiry into the human and moral significance of developments in biomedical and behavioral science and technology.
2. To explore specific ethical and policy questions related to these developments.
3. To provide a forum for a national discussion of bioethical issues.
4. To facilitate a greater understanding of bioethical issues.
5. To explore possibilities for useful international collaboration on bioethical issues."[88]

As with its predecessors, the council has carried out studies and issued reports on a number of subjects, with issues of stem cell research receiving perhaps greater attention than in earlier years. In July 2002, for example, it published *Human Cloning and Human Dignity: An Ethical Inquiry*, an effort to explore issues involved in cloning practices for the purposes of producing humans and for producing embryos for research purposes. In the conclusion to its report, the commission noted that its members had been unanimously opposed to the use of cloning for human reproductive purposes, but had been divided almost equally between the approval of and opposition to cloning for the purpose of medical therapeutic purposes (seven members favored the practice, seven opposed the practice, and three favored a moratorium on the practice).

In January 2004, the commission produced another major report on stem cell research, *Monitoring Stem Cell Research*, a report with three major foci: (1) the science and technology of stem cell research, with a review of recent developments in the field; (2) the legal and policy history and current status of stem cell research in the United States; and (3) ethical issues related to the practice of stem cell research. The report makes no policy recommendations and is most useful for the summary it provides of the science, ethics, and legal status of stem cell research. Also of special interest are 10 appendices consisting of commissioned papers on a number of specific topics, such as current progress in human embryonic stem cell research, adult stem cells, the biology of nuclear cloning and the potential of embryonic stem cells for transplantation therapy, and stem cells and tissue regeneration.

The council's most recent report, *Alternative Sources of Pluripotent Stem Cells: A White Paper*, was issued in May 2005. It was prepared in response to the growing debate over the use of embryonic stem cells in research as an attempt to suggest sources of pluripotent stem cells that do not involve the death of a fertilized egg, blastocyst, or embryo. The four sections of the report consider the collection of pluripotent stem cells from organismically dead embryos, via blastomere extraction from living embryos, from biological artifacts, and by somatic cell dedifferentiation.[89]

The dearth of federal law on stem cell research does not reflect efforts to produce legislation on this topic. In fact, since President George W. Bush's August 9, 2001, statement outlining federal policy on stem cell research, a number of bills in both the Senate and the House of Representatives have been introduced to clarify the nation's policy on SCR. The goal of some bills has been to reverse the president's policy on stem cell research, and the goal of others, to solidify that policy. Space does not permit a detailed analysis of these bills, but some of the legislation introduced into the 109th Congress includes the following:

- H.R. 162 Stem Cell Replenishment Act of 2005 (Rep. Juanita Millender-McDonald (D-Calif.); permitting the use of federal funds to support embryonic stem cell research)
- H.R. 810 Stem Cell Research Enhancement Act of 2005 (Rep. Mike Castle (R-Del.); permitting the use of federal funds to support embryonic stem cell research)
- S. 471 Stem Cell Research Enhancement Act of 2005 (Sen. Arlen Specter (R-Pa.); permitting the use of federal funds to support embryonic stem cell research)
- H.R. 1650 Stem Cell Research Investment Act of 2005 (Rep. Nancy Johnson (R-Conn.); granting tax credits and other financial incentives for the conduct of stem cell research)
- H.R. 2520 Stem Cell Therapeutic and Research Act of 2005 (Rep. Christopher Smith (R-N.J.): providing for the creation of a cord blood bank, a repository for blood collected from fetal umbilical cords)
- S. 659 Human Chimera Prohibition Act of 2005 (Sen. Sam Brownback (R-Kans.); prohibiting the production of chimeras [combinations of human and nonhuman components in a fertilized egg] in research)
- S. 1373 Human Chimera Prohibition Act of 2005 (Sen. Sam Brownback (R-Kans.); similar to S. 659)
- S. 876 Human Cloning Ban and Stem Cell Research Protection Act of 2005 (Sen. Orin Hatch (R-Utah); prohibiting cloning for the purpose of

human reproduction and establishing conditions for other types of cloning)

- H.R. 1822 Human Cloning Ban and Stem Cell Research Protection Act of 2005 (Rep. Mary Bono (R-Calif.); parallel legislation to S. 876)
- H.R. 222 Human Cloning Research Prohibition Act (Rep. Cliff Stearns (R-Fla.); prohibiting cloning for the purpose of human reproduction and outlining purposes for which somatic cell nuclear transfer can be used)
- S. 658 Human Cloning Prohibition Act of 2005 (Sen. Sam Brownback (R-Kans.); comprehensive ban on all types of human cloning)
- H.R. 1357 Human Cloning Prohibition Act of 2005 (Rep. Dave Weldon (R-Fla.); similar to S. 658)
- H.R. 776 Sanctity of Life Act of 2005 (Rep. Ronald Paul (R-Tex.); making it federal policy that life begins at conception)

In most instances, these bills, like their predecessors in earlier sessions of the Congress, have not progressed very far in the legislative process, usually not surviving the committees to which they were assigned. The exceptions thus far have been H.R. 810, which passed the House of Representatives on a vote of 238-194 on May 24, 2005, and S. 471, which passed the Senate on a vote of 63-37 on July 18, 2006. President Bush vetoed the legislation, however, and the House of Representatives was unable to override the veto, so the bills died.

State Actions

As is to be expected, various states have adopted a variety of positions on stem cell research, some adopting legislation essentially consistent with federal policies, and some choosing to become more supportive of a greater variety of stem cell research projects. As of early 2006, 16 states had specific restrictions on any form of research conducted on human embryos and fetuses.[90] The same number of states, but not necessarily the same states, prohibit research on fetuses or embryos produced from sources other than abortion. In a report issued in 2006, the National Conference of State Legislatures lists nine states—Arizona, Arkansas, Illinois, Indiana, Massachusetts, Michigan, Nebraska, North Dakota, and South Dakota—that have laws prohibiting research on aborted fetuses or fetuses resulting from any source other than abortion. Another 16 states restrict the use of either aborted fetuses or fetuses obtained from some other source, although not both.[91]

In contrast to the decisions made in these states, a number of other states had decided to respond to the lack of federal funds for embryonic stem cell research by encouraging and supporting such research using their own funds. The first two states to move in this direction were California and New Jersey. In California, the leading proponent of embryonic stem cell research has been State Senator Deborah Ortiz (D-Sacramento), who has sponsored a number of bills in the state senate promoting such research. One of those bills, SB 253, legalized research on the production and use of embryonic stem cells. The bill also established standards for the donation and use of human embryos and fetal tissue, created a permanent biomedical advisory committee, and prohibited cloning for reproductive purposes. When this bill was signed into law in 2002, it made California the first state in the nation to formally promote and advance stem cell research with public funds.

SB 253 was only the first step in the promotion of stem cell research in California, however. When the state legislature rejected another bill sponsored by Ortiz creating a state agency to promote and finance stem cell research, supporters of SCR turned to the initiative process to achieve the same objective. They drafted Proposition 71, an initiative for the creation of a California Institute for Regenerative Medicine, whose purpose it would be "to regulate stem cell research and provide funding, through grants and loans, for such research and research facilities."[92] The initiative called for funding of the institute with $3 billion in general obligation bonds over a 10-year period. It appeared on the November 2004 ballot and was approved by voters by a 59 to 41 percent margin. By late 2005, the institute had begun the organization process and had chosen San Francisco as the site for its headquarters.

New Jersey became the second state to adopt legislation legalizing stem cell research in the state and developing a system for the promotion and financial support of such research. On January 2, 2004, the state senate and general assembly passed an act creating a Stem Cell Institute of New Jersey, dedicated to basic research on stem cells and their applications in therapeutic medicine. Funding for the institute was set at $380 million, $150 million in a direct grant from the state and $230 million to come from public issue bonds. The institute was to be operated jointly by the University of Medicine and Dentistry of New Jersey and Rutgers University.

Proposals for the support and funding of stem cell research in other states are also going forward. In May 2005, for example, Massachusetts became the third state to adopt pro-SCR legislation when both houses of the state legislature overrode a veto by governor Mitt Romney and adopted a far-reaching program for the support of stem cell research in the state. A month later, similar legislation was passed by the Connecticut legislature and signed by governor Jodi Rell. And in July, Illinois governor Rod Blagojevich issued Executive Order 6 allocating $10 million in state funds to support stem cell research in the state.

STEM CELL RESEARCH: PROGRESS AND PROMISE

Stem cell research is still in its infancy. Some of the most important basic discoveries in the field go back no more than two decades, some even less than that. But hopes for the medical potential of this research is high. Proponents suggest that stem cell research may provide cures to any number of diseases and disorders that are currently intractable to other forms of treatment. Stem cells, they argue, may produce revolutionary changes in medicine seen only rarely in human history.

But such advances are not likely to come easily. Many technical problems remain to be solved in the production, extraction, and use of stem cells. And a number of fundamental ethical issues remain to be resolved. When researchers work with human embryos and fetuses, are they destroying human life and, if so, is that sacrifice justifiable on the basis of the medical benefits that may result? Answers to those questions have eluded scientists, ethicists, religious leaders, politicians, and ordinary citizens for more than two decades. And they are likely to escape an easy solution for some time to come.

---

[1] C. Ward Kischer, American Bioethics Advisory Commission. "Research on Stem Cells," Available online. URL: http://www.all.org/abac/cwk003.htm. Downloaded on June 27, 2005.

[2] See, for example, Daniel R. Marshak, David Gottlieb, and Richard L Gardner, "Introduction: Stem Cell Biology," in Daniel R, Marshak, David Gottlieb, and Richard L. Gardner, eds. *Stem Cell Biology*. Cold Spring Harbor, N.Y.: Cold Spring Harbor Laboratory Press, 2001, pp. 1–2, and Jane Maienschein. *Whose View of Life?: Embryos, Cloning, and Stem Cells*. Cambridge, Mass.: Harvard University Press, 2003, p. 253.

[3] "What Medical Advances Has the Jackson Laboratory Enabled?" The Jackson Laboratory. Available online. URL: http://www.jax.org/mission/tjl_advances.html. Downloaded on June 27, 2005, and Ann B. Parson. *The Proteus Effect*. Washington, D.C.: Joseph Henry Press, 2004, p. 61.

[4] As cited in Scott F. Gilbert. *Developmental Biology*, 6th edition. Textbook available online. URL: http://www.ncbi.nlm.nih.gov/books/bv.fcgi?rid=dbio.section.4360. Downloaded on June 27, 2005.

[5] Andrew Pollack, "Missing Limb? Salamander May Have Answer," *New York Times*, September 24, 2002, p. F1.

[6] For an excellent review of Trembley's work, see Parson, Chapter 1.

[7] Cited in "Epigenesis/Preformation." Christian Hubert. Available online. URL: http://www.christianhubert.com/hypertext/Epigenesls_Preformation.html. Downloaded on June 27, 2005.

[8] Alexander Maximow, "Der Lymphozyt als gemeinsame Stammzelle der verschiedenen Blutelemente in der embryonalen Entwicklung und im postfetalen Leben der Säugetiere," *Folia Haematologia* (Leipzig), vol. 8. 1909, pp. 125–141. Cited and translated in Theodor M. Fliedner, "Prologue to Characteristics and Potentials of Blood Stem Cells," *Stem Cells*, vol. 16, no. 6, November 1998, p. 357.

[9] The paper was L. Siminovitch, E. A. McCulloch, and J. E. Till, "The Distribution of Colony-Forming Cells Among Spleen Colonies," *Journal of Cellular and Comparative Physiology*, vol. 62, December 1963, pp. 327–336. The quotation is from Parson, p. 61.

[10] "Canadians Till and McCulloch Proved Existence of Stem Cells." Stem Cell Network. Available online. URL: http://www.stemcellnetwork.ca/news/articles.php?id=539. Last modified August 25, 2005.

[11] Isidore Geoffrey Saint-Hilaire. *Histoire générate et particulière des anomalies de I'organisation chez l'homme et les animaux. Des monstruosités, des varietés et vices de conformation, ou traité de tératologie.* Paris: J-B. Baillière, 1832, p. 19, as quoted and translated in Melinda Cooper, "Regenerative Medicine: Stem Cells and the Science of Monstrosity," *Journal of Medical Humanities*, vol. 30, no. 1, June 2004, p. 15.

[12] Cooper, p. 17.

[13] Camille Dareste. *Recherches sur la production artificielle des monstruosités ou, essais de tératogénie expérimentale*, 2nd ed. Paris: C Reinwald, 1891, p. 24, as quoted and translated in Cooper, p. 17.

[14] L. C. Stevens. "The Development of Transplantable Teratocarcinomas from Intratesticular Grafts of Pre- and Postimplantation Mouse Embryos." *Developmental Biology*, vol. 21, no. 3, March 1970, pp. 364–382.

[15] Stanley Shostak. *Becoming Immortal: Combining Cloning and Stem Cell Therapy.* New York: State University of New York Press, 2001, p. 182, as quoted in Melinda Cooper, "Regenerative Medicine: Stem Cells and the Science of Monstrosity," *Journal of Medical Humanities*, vol. 30, no. 1, June 2004, p. 17.

[16] B. Mintz and K. Illmensee, "Normally Genetically Mosaic Mice Produced from Malignant Tumor Cells," *Proceedings of the National Academy of Sciences*, vol. 72, no. 9, September 1975, pp. 3585–3589, and R. L. Brinster, "The Effect of Cells Transferred into the Mouse Blastocyst on Subsequent Development," *The Journal of Experimental Medicine*, vol. 140, no. 4, October 1, 1974, pp. 1049–1056.

[17] M. J. Evans and M. H. Kaufman, "Establishment in Culture of Pluripotential Cells from Mouse Embryos," *Nature*, vol. 292, no. 5819, July 9, 1981, pp. 154–156.

[18] G. R. Martin, "Isolation of a Pluripotent Cell Line from Early Mouse Embryos Cultured in Medium Conditioned by Teratocarcinoma Stem Cells," *Proceedings of the National Academy of Sciences*, vol. 78, no. 12, December 1981, pp. 7634–7638.

[19] Centers for Disease Control and Prevention, National Center for Chronic Disease Prevention and Health Promotion, Division of Reproductive Health. *2002 Assisted Reproductive Technology Success Rates.* Atlanta, Ga.: Centers for Disease Control and Prevention, December 2004, p. 11.

[20] "Japanese In-Vitro Fertilization Births Top 100,000," *Mainichi Shimbun*, January 27, 2005. Available online. URL: http://66.102.7.104/search?q=cache:Wl-QXaoLckkJ:202.221.31.68/news/archive/200501/27/20050127p2aOOmOdm010000c.html+%22in+vitro+fertilizatlon+births%22+japan&hl=en. Posted on January 27, 2005.

[21] David I. Hoffman, et al., "Cryopreserved Embryos in the United States and Their Availability for Research," *Fertility and Sterility*, vol. 79, no. 5, May 2003, p. 1063.

[22] James A. Thomson, et al., "Embryonic Stem Cell Lines Derived from Human Blastocysts," *Science*, vol. 282, no. 5391, November 6, 1998, pp. 1145–1147.

[23] Michael J. Shamblott, et al., "Derivation of Pluripotent Stem Cells from Cultured Human Primordial Germ Cells," *Proceedings of the National Academy of Sciences*, vol. 95, no. 23, November 10, 1998, pp. 13726–13731.

[24] Hwang, Woo Suk, et al., "Patient-specific Embryonic Stem Cells Derived from Human SCNT Blastocysts," *Science*, vol. 308, no. 5729, June 17, 2005, pp. 1777–1783.

[25] *ASRM Bulletin*, vol. 7, no. 16, May 19, 2005. Also available online. URL: http://www.asrm.org/Washington/Bulletins/vol7no16.html.

[26] LifeSite. "Korean Cloning Doctor Creates and Kills Clones from Patients with Spinal Cord Injuries." Available online. URL: http://www.lifesite.net/ldn/2005/may/05052009.html. Posted on May 20, 2005.

[27] This section provides a brief, nontechnical summary of stem cell science. One of the best general summaries of stem cell research for the general reader is the one provided by the National Institutes of Health at its web site on stem cells. Available online. URL: http://stemcells.nih.gov/info/basics/. Downloaded on July 6, 2005.

[28] For an excellent review of the current status of this research, see Martin Raff, "Adult Stem Cell Plasticity: Fact or Artifact?" *Annual Review of Cell and Developmental Biology*, vol. 19, November 2003, pp. 1–22.

[29] Center for Regenerative Medicine [of the Medical University of South Carolina]. Available online. Centers. URL: http://research.musc.edu/bp/centers_regmed.html. Revised date August 2005.

[30] National Institutes of Health. *Stem Cells: Scientific Progress and Future Research Directions.* Washington, D.C.: Department of Health and Human Services, June 2001.

[31] Nadya, Lumelsky, et al., "Differentiation of Embryonic Stem Cells to Insulin-Secreting Structures Similar to Pancreatic Islets. *Science*, vol. 292, no. 5520, May 18, 2001, pp. 1389–1394.

[32] National Institutes of Health, *Stem Cells*, p. ES-10.

[33] "Science and Nature." PollingReport.com. Available online. URL: http://www.pollingreport.com/science.htm. Downloaded on July 8, 2005. This web site summarizes the results of a number of public opinion polls on various aspects of stem cell research and other topics. For polls conducted in 2005, support for embryonic stem cell research ranged from 58 percent to 71 percent.

[34] For an excellent overview of this issue, see Jane Maienschein, "What's in a Name: Embryos, Clones, and Stem Cells," *American Journal of Bioethics*, vol. 2, no. 1, Winter 2002, pp. 12–19.

[35] "Human Stem Cell Research." Religious Tolerance.org. Available online. URL: http://www.religioustolerance.org/res_stem3.htm. Latest update November 8, 2001.

[36] Clifford Grobstein, "External Human Fertilization," *Scientific American*, vol. 240, no. 6, June 1979, pp. 57–67. For one application of the term pre-embryo to stem cell research issues, see Richard McCormick, "Who or What is the Preembryo?," *Kennedy Institute of Ethics Journal*, vol. 1, no. 1, March 1991, pp. 1–15.

[37] C. Ward Kischer, "The Big Lie in Human Embryology: The Case of the Preembryo." lifeissues.net. Available online. URL: http://www.lifeissues.net/writers/kisc/kisc_11bigliepreembryo.html#a3. Downloaded on July 9, 2005.

[38] See, for example, "What Is 'Bioethics'?" lifeissues.net. Available online. URL: http://www.lifeissues.net/writers/irv/irv_36whatisbioethics10.html. Downloaded on July 9, 2005.

[39] Harold J. Morowitz and James S. Trefil. *The Facts of Life: Science and the Abortion Controversy*. New York: Oxford University Press, 1994, as cited in Scott F. Gilbert, "When Does Life Begin." Textbook available online. URL: http://www.devbio.com/article.php?id=162. Downloaded on July 9, 2005.

[40] Fritz Baumgartner. "Life Begins at the Beginning (A Doctor Gives the Scientific Facts on When Life Begins)." Pro-Life America. Available online. URL: http://www.prolife.com/life_begins.html. Posted April 12, 2005.

[41] For more on this topic, see "When Does Life Begin?" in Scott F. Gilbert. *DevBio, A Companion to Developmental Biology*, 7th edition. Textbook available online. URL: http://www.devbio.com/article.php?id=162. Downloaded on July 9, 2005.

[42] "Abortion and Catholic Thought: The Little-Told History." Catholics for Choice. Available online. URL: http://www.catholicsforchoice.org/nobandwidth/English/cathwomen/abortiondecision. htm. Downloaded on July 9, 2005.

[43] Ioannes Paulus PP. II. *Evangelium Vitae:* To the Bishops Priests and Deacons Men and Women religious lay Faithful and all People of Good Will on the Value and Inviolability of Human Life, p. 61. Available online. URL: http://www.vatican.va/holy_father/john_paul_ii/encyclicals/documents/hf_jp-ii_enc_25031995_evangelium-vitae_en.html#-10. Promulgated on March 25, 1995.

[44] *Ibid.*, p. 45.

[45] *Ibid.*, p. 63.

[46] Margaret A. Farley, "Roman Catholic Views on Research Involving Human Embryonic Stem Cells," *Ethical Issues in Human Stem Cell Research, Vol. III: Religious Perspectives*. Washington, D.C.: National Bioethics Advisory Commission, June 2000, p. D-4.

[47] Thomas A. Shannon. "Ethical Issues in Stem Cell Therapy from the Micro to the Macro." WPI Transformations. Available online. URL: http://wpi.edu/News/Transformations/2003Spring/stemcell.html. Downloaded on July 11, 2005.

[48] *Ibid.*

[49] John J. Conley, S.J., "Delayed Animation: An Ambiguity and Its Abuses." University Faculty for Life. Available online. URL: http://uffl.org/vol12/conley12.pdf. Downloaded on July 11, 2005.

[50] "Orthodox Church & Stem Cell Research." Orthodox Church in America. Available online. URL: http://www.oca.org/QA.asp?ID=68&SID=3. Downloaded on July 11, 2005.

[51] "30th Meeting of the National Bioethics Advisory Commission: Religious Views on Research Involving Human Embryonic Stem Cells." Georgetown University. Available online. URL: http://www.georgetown.edu/research/nrcbl/nbac/transcripts/may99/may_7.pdf. Written on May 7, 1999.

[52] "Statement on Human Stem Cell Research." Faith and Family Issues. Available online. URL: http://sites.silaspartners.com/partner/Article_Display_Page/0,,PTID314166|CHID599214|CIID1890076,00.html. Downloaded on July 11, 2005.

[53] Convention Proceedings, 61st Regular Convention, The Lutheran Church—Missouri Synod, St. Louis, Missouri, July 14–20, 2001, p. 162. Also available online. URL: http://www.lcms.org/graphics/assets/media/2004%20Convention/2001convproceedings.pdf.

[54] "Support for Federally Funded Research on Embryonic Stem Cells." General Synod, United Church of Christ. Available online. URL: http://www.ucc.org/synod/resolutions/res30.htm. Downloaded on July 11, 2005.

[55] "The Archives of the Episcopal Church: Acts of Convention, 1976–2003." The Archives of the Episcopal Church. Available online. URL: http://www.episcopalarchives.org/cgi-bin/acts/acts_resolution.pl?resolution=2003-A014. Downloaded on July 11, 2005.

[56] Elliott N. Dorff, "Testimony of Rabbi Elliot N. Dorff, Ph.D.," in Michael Ruse and Christopher Pynes, eds. *The Stem Cell Controversy: Debating the Issues.* Amherst, N.Y.: Prometheus Books, 2003, p. 197.

[57] "Resolution on Stem Cell Research." Union for Reform Judaism. Available online. URL: http://urj.us/cgi-bin/resodisp.pl?file=stemcell&year=2003N. Downloaded on July 11, 2005.

[58] "Action Alert: Stem Cell Research Enhancement Act of 2005 (H.R. 810)." Institute for Public Affairs. Available online. URL: http://www.ou.org/public/actionalerts/2005/n1.htm. Downloaded on July 11, 2005.

[59] Dr. Muzammil H. Siddiqi. "Stem Cell Research: An Islamic Perspective." Crescentlife.com. Available online. URL: http://www.crescentlife.com/wellness/stem_cell_research_islamic_prespective.htm. Downloaded on July 11, 2005.

[60] C. Ben Mitchell. "NIH, Stem Cells, and Moral Guilt." The Center for Bioethics. Available online. URL: http://www.cbhd.org/resources/stemcells/mitchell_2000-08-24.htm. Downloaded on July 12, 2005.

[61] John A. Robertson, "Ethics and Policy in Embryonic Stem Cell Research," in Michael Ruse and Christopher Pynes, eds. *The Stem Cell Controversy: Debating the Issues.* Amherst, N.Y.: Prometheus Books, 2003, pp. 121–150.

[62] Carolyn Gargaro. "Bush and Embryonic Stem Cell Research." Rightgirl! Available online. URL: http://rightgirl.com/carolyn/bstemcell.shtml. Posted August 13, 2001.

[63] National Bioethics Advisory Commission. *Ethical Issues in Human Stem Cell Research.* Washington, D.C.: National Technical Information Service, U.S. Department of Commerce, September 1999, vol. 1, p. 50.

[64] Heather Johnson Kukla, "Embryonic Stem Cell Research: An Ethical Justification," *Georgetown Law Journal*, vol. 90, no. 2, January 2002, pp. 503–543.

[65] David A. Prentice, "Adult Stem Cells," in *Monitoring Stem Cell Research: A Report of the President's Council on Bioethics.* Washington, D.C.: Government Printing Office, 2004, Appendix K, pp. 309–346.

[66] *Ibid.*, p. 331.

[67] "Adult Stem Cell Research." Eureka Alert. Available online. URL: http://www.eurekalert.org/pub_releases/2003–09/nymc-asc091503.php. Posted on September 18, 2003.

[68] "Regenerative Chemical Turns Muscle Cells into Stem Cells, Say Scientists at The Scripps Research Institute." The Scripps Research Institute. Available online. URL: http://www.scripps.edu/news/press/122203.html. Posted on December 22, 2003.

[69] See, for example, "Potential Seen in Adult Stem Cells." CNN.com International. Available online. URL: http://edition.cnn.com/2005/TECH/science/03/21/australia.stemcell. Posted on March 22, 2005.

[70] Wolfgang Lillge. "The Case for Adult Stem Cell Research," *21st Century Science and Technology Magazine*, Winter 2001–2002. Also available online. URL: http://www.21stcenturysciencetech.com/articles/winter01/stem_cell.html.

[71] Gillian M. Beattie, et al., "Activin A Maintains Pluripotency of Human Embryonic Stem Cells in the Absence of Feeder Layers," *Stem Cells*, vol. 23, no. 4, April 2005, pp. 489–495.

[72] Christoper J. Flaim, Sangeeta Bhatia, and Shu Chien, "An Extracellular Matrix Microarray for Probing Cellular Differentiation," *Nature Methods*, vol. 2, no. 2, January 21, 2005, pp. 119–125.

[73] Xuejun Li, et al., "Specification of Motoneurons from Human Embryonic Stem Cells," *Nature Biotechnology*, vol. 23, no. 2, February 1, 2005, pp. 215–221.

[74] Hans S. Keirstead, et al., "Human Embryonic Stem Cell-Derived Oligodendrocyte Progenitor Cell Transplants Remyelinate and Restore Locomotion after Spinal Cord Injury", *Journal of Neuroscience*, vol. 25, no. 19, May 11, 2005, pp. 4694–4705.

[75] "What's New." Stem Cell Research Foundation. Available online. URL: http://www.stemcellresearchfoundation.org/WhatsNew/May_2005.htm#1. Posted May 2005.

[76] "Report of the Committee of Inquiry into Human Fertilisation and Embryology." London: Her Majesty's Stationery Office, July 1984. The full report is available online at http://www.bopcris.ac.uk/imgall/ref21165_1_1.html.

[77] The Donaldson report is available in full online at http://www.dh.gov.uk/assetRoot/04/06/50/85/04065085.pdf.

[78] "UK Cloning Projects in Disarray after Prolife Alliance High Court Victory." ProLife. Available online. URL: http://www.prolife.org.uk/docstatic.asp?id=VICTORY.htm&se=2&st=6. Downloaded on July 18, 2005.

[79] "Summary of Conclusions and Recommendations." The United Kingdom Parliament. Available online. URL: http://www.parliament.the-stationery-office.co.uk/pa/ld200102/Idselect/Idstem/83/8310.htm. Downloaded on July 18, 2005.

[80] This information is accurate as of late 2003. See "Regulations in European Member States Regarding Human Embryonic Stem Cell Research." International Society for Stem Cell Research. Available online. URL: http://www.isscr.org/scientists/legislative.htm. Posted on October 17, 2003.

[81] Heather Tomlinson and David Adam. "China Leads in Stem Cell Research," *The Hindu*, January 25, 2005. Also available online. URL: http://www.hindu.com/2005/01/25/stories/2005012500721500.htm.

[82] For more information on the Tuskegee study, see "Tuskegee Syphilis Study," Centers for Disease Control and Prevention. Available online. URL: http://www.cdc.gov/nchstp/od/tuskegee. Posted on May 23, 2005.

[83] The report is available in the Federal Register for 1979, volume 44, pp. 35033–35058.

[84] "Former Bioethics Commissions." The President's Council on Bioethics. Available online. URL: http://www.bioethics.gov/reports/past_commissions. Downloaded on July 20, 2005.

[85] "White Paper: Alternative Sources of Pluripotent Stem Cells." President's Council on Bioethics. Available online. URL: http://www.bioethics.gov/reports/white_paper/text.html#_edn10. Posted in May 2005.

[86] Executive Order 12975 of October 3, 1995. Protection of Human Research Subjects and Creation of National Bioethics Advisory Commission. Available online. URL: http://www.georgetown.edu/research/nrcbl/nbac/about/eo12975.htm. Downloaded on July 20, 2005.

[87] Judith A. Johnson and Brian A. Jackson. "Stem Cell Research," CRS Report for Congress, September 19, 2000, p. CRS-3. Also available online. URL: http://www.law.umaryland.edu/marshall/crsreports/crsdocuments/RS20523.pdf.

[88] The President's Council on Bioethics. The President's Council on Bioethics. Available online. URL: http://permanent.access.gpo.gov/lps21821/www.bioethics.gov. Downloaded on July 20, 2005.

[89] The report is available online at http://www.bioethics.gov/reports/white_paper/alternative_sources_white_paper.pdf. Downloaded on July 20, 2005.

[90] A chart providing current information on state policies on stem cell research is available at "State Embryonic and Fetal Research Laws." National Conference of State Legislatures. Available online. URL: http://www.ncsl.org/programs/health/genetics/embfet.htm. Downloaded on July 21, 2005.

[91] Alissa Johnson. "State Embryonic and Fetal Research Laws." National Conference of State Legislatures. Available online. URL: http://www.ncsl.org/programs/health/genetics/embfet.htm. Downloaded on March 14, 2006.

[92] "Proposition 71: Stem Cell Research. Funding. Bonds. Initiative Constitutional Amendment and Statute." Attorney General, State of California. Available online. URL: http://www.ss.ca.gov/elections/bp_nov04/prop_71_entire.pdf. Downloaded on July 21, 2005.

CHAPTER 2

THE LAW AND STEM CELL RESEARCH

This chapter describes laws, legislative actions, administration rulings, and legal decisions relating to stem cell research conducted in the United States. Extracts from some of these laws, court decisions, and other legal documents appear in the appendices.

LAWS AND ADMINISTRATIVE ACTIONS

Stem cell research has been of public concern for only a relatively short period of time. Thus far, few laws at the federal or state level dealing specifically with stem cell research have been passed. Most SCR regulations occur in legislation passed for somewhat different purposes, such as rules covering human experimentation, or as administrative orders. The documents listed below are some of the most important of these laws and regulations.

PROTECTION OF HUMAN SUBJECTS, 45 CFR 46.208(A)(2) (1974) AND 42 USC §289-1 (1993)

Two of the very few federal laws that regulate stem cell research are to be found in the Code of Federal Regulations (CFR), Chapter 45, Section 46, Subsection 208(a)(2) and the United States Code (USC), Title 42, Chapter 6A, Subchapter III, Part H, paragraph 289g-1, both dealing with the protection of human fetuses. The former section permits research or experimentation on a human fetus under only two circumstances: (1) if the health of the fetus itself is at risk and (2) if experimentation on the fetus will cause minimal harm to it and will produce medical knowledge that cannot be obtained by any other means. The definition of a fetus used in this section of the code is "the product of conception from the time of implantation (as evidenced by any of the presumptive signs of pregnancy, such as missed menses, or a medically acceptable pregnancy test)," and thus includes the structure usually called an embryo also. A number of writers have commented on this somewhat unusual (or, as one observer has written, "rather bizarre"[1]) definition of *fetus*, which differs significantly from that used by almost all (if not all) embryologists as referring to the structure that develops after about the eighth week of pregnancy. In any case, the definition is still part of the U.S. Code in regulations dealing with experimentation on what most people would now call the embryo, as well as the fetus into which it later develops.

45 CFR 46.208(a)(2) grew out of the National Research Act, passed by the U.S. Congress in 1974 after a fairly lengthy debate over the ethics of experimentation on human subjects. One of the major motivations for that legislation was revelations of abuses that had occurred in the use of humans for experiments in the now infamous Tuskegee Syphilis Study, a project sponsored by the U.S. Public Health Service from 1932 to 1972 in Macon County, Alabama. In that study, African-American subjects were intentionally given syphilis to learn more about the nature and course of the disease. Treatment that would have cured the subjects was withheld, and many died or were permanently maimed as the disease progressed.

During the congressional debate over various bills designed to deal with the ethical problems of human experimentation, information was released about the use of whole, live fetuses in various research projects funded by the National Institutes of Health. This information prompted calls for inclusion of protection for "unborn children" in any legislation on human experimentation. It was this concern that led to the eventual adoption of the language of 45 CFR 46.208(a)(2), which remains one of the few legislative guidelines for stem cell research in the United States.

The relevant section of the U.S. Code on embryonic stem cell research, 42 USC §289-1, is titled "Research on transplantation of fetal tissue." It authorizes the secretary of Health and Human Services to use fetal tissue for therapeutic purposes "regardless of whether the tissue is obtained pursuant to a spontaneous or induced abortion or pursuant to a stillbirth." The section defines human fetal tissue as "tissue or cells obtained from a dead human embryo or fetus after a spontaneous or induced abortion, or after a stillbirth." It provides the conditions under which such tissue may be obtained (always with the consent of the donor) and prohibits the payment for any fetal tissue used in therapeutic procedures.

The legislative origin of 42 USC §289-1 is the NIH Revitalization Act enacted on June 10, 1993. The act was a far-reaching bill designed to increase federal support for research in virtually all of the fields in which NIH has research interest. New programs were outlined for the National Institute of Cancer; National Heart, Lung, and Blood Institute; National Institute of Diabetes and Digestive and Kidney Diseases; National Institute of Arthritis and Musculoskeletal and Skin Diseases; National Institute on Aging; National Institute of Allergy and Infectious Diseases; National Institute of Child Health and Human Development; National Institute of Neurological Disorders and Stroke; National Institute of Environmental Health Sciences; the National Eye Institute; and the National Library of Medicine. Interest in research using transplanted fetal tissue was emphasized by placement of that topic at the very beginning of the bill, following a section on general provisions governing proposals for biomedical and behavioral research. The purpose of the section, freely and openly recognized by the bill's sponsors, was to remove the existing moratorium on fetal research that had been in place since March 22, 1988. Even the most staunch conservatives in the U.S. Senate expressed their support for the bill and its fetal transplantation section. For example, Senator Strom Thurmond (R-S.C.), one of Congress's most conservative members, was quoted as saying that "This is not an abortion issue. It is a research issue. It is not about taking lives. It is about saving and improving lives."[2] Although the bill was eventually passed, signed by President Bill Clinton on June 10, 1993, and codified in the U.S. Code at 42 USC §289-1, its provisions have been largely ignored. Within 18 months of its adoption, a new president, George W. Bush, had taken office and announced a new and more restrictive approach to the use of fetal materials in research.

A final note about the NIH Revitalization Act: In section 121(c) of Public Law 103-43, the provisions of President Clinton's Executive Order 12806 of May 19, 1992, establishing a human fetal tissue bank supported by the U.S. government, was nullified and declared not to have any legal effect.

DICKEY AMENDMENT OF 1995

The Dickey amendment was first proposed in 1995 by Representative Jay Dickey (R-Ark.) as a rider to the appropriations bill for the Department of Health and Human Services (HHS) for fiscal year 1997. It has been renewed in much the same form every year since. The amendment prohibits HHS from using appropriated funds for the creation of human embryos for research or for research that makes use of human embryos that have been destroyed. Although the language of the rider varies slightly from year to year, its overall intent remains essentially the same. The wording of the rider for fiscal year 2005, for example, prohibits:

> *(1) the creation of a human embryo or embryos for research purposes; or (2) research in which a human embryo or embryos are destroyed, discarded, or knowingly subjected to risk of injury or death greater than that allowed for research on fetuses in utero under 45 CFR 46.208(a)(2) and section 498(b) of the Public Health Service Act (42 U.S.C. 289g(b)).*[3]

The amendment also defines the terms human embryo or embryos as "any organism, not protected as a human subject under 45 CFR 46 as of the date of the enactment of this Act, that is derived by fertilization, parthenogenesis, cloning, or any other means from one or more human gametes or human diploid cells."

The precise wording of the Dickey amendment for the fiscal years 1996 through 2005 can be found in the following public laws: 1996: 104-99; 1997: 104-208; 1998: 105-78; 1999: 105-277; 2000: 106-113; 2001: 106-554; 2002: 107-116; 2003: 108-7, 2004: 108-199; and 2005: 108-447.

Lacking any other federal regulation on embryonic stem cell research, the Dickey amendment has been one of the major cornerstones of federal policy on this subject. Although it was written to restrict the actions of one department only, Health and Human Services, President Bill Clinton extended the ban to all federal agencies in an executive order issued in March 1997.

One of the most serious challenges to the Dickey regulations occurred during the Clinton administration. In late 1998, officials at the National Institutes of Health asked Harriet Rabb, General Counsel for HHS, for a legal opinion as to precisely what the Dickey amendment meant for the funding of stem cell research. Rabb issued her opinion in January 1999, ruling that federal funds could not be used for the *creation* of human embryos for research, but they could be used for research on stem cells that had been removed from human embryos produced through private funding.

Rabb's ruling became the basis of new guidelines on stem cell research published in August 2000, shortly before the presidential election in which George W. Bush defeated Al Gore. Given the change in presidential administrations and their philosophies on stem cell research, the newly published guidelines were never put into effect and were eventually supplanted by President Bush's August 2001 speech on stem cell research.

GEORGE W. BUSH'S PRESIDENTIAL ADDRESS ON STEM CELL RESEARCH (AUGUST 9, 2001)

During his campaign for the presidency leading up to the 2000 election, George W. Bush indicated that he was opposed to any form of stem cell research in which human embryos were created and then destroyed for the production of stem cells. Shortly after he took office in January 2001, Bush put a hold on a government program established to accept and review grant proposals for embryonic stem cell research until his own administration could review the state of affairs in the field. The review committee issued its report on July 13, 2001, recommending that the government support both embryonic and adult stem cell research. Only a month later, Bush made a televised speech from his ranch in Crawford, Texas, to announce his policy on stem cell research.

In that speech, Bush acknowledged the potential value of using embryonic stem cell in research for the cure of a number of disabling diseases and disorders. He also reviewed the ethical issues involved in conducting such research. One of the most powerful factors in the decision he eventually made, Bush said, was his belief that embryos are living human beings and should not be killed for any reason, including possible medical benefits that might arise from stem cell research.

His decision on this contentious issue was that the U.S. government under his administration would be allowed to fund research on stem cell lines that were already in existence, in most cases the "left-over" embryos produced for the purpose of in vitro fertilization, but that funding would not be permitted for the purpose of creating new embryos for research. Bush estimated that a total of 60 stem cell lines—later increased to 78 stem cell lines—met his criterion and were eligible for funding support. The actual number of embryonic lines in good enough condition for research has been something of a matter of debate ever since. In any case, recalling science fiction novelist Aldous Huxley's description of baby "hatcheries" in his novel *Brave New World*, Bush rejected the possibility of allowing the federal funding of any new stem cell lines. In conclusion, he announced the creation of a new President's Bioethical Advisory Commission "to recommend appropriate guidelines and regulations, and to consider all of the medical and ethical ramifications of biomedical innovation."

The president's decision on the funding of stem cell research turned out to satisfy only a relatively small fraction of the general public. Many of his supporters were disappointed that he had decided to permit *any* embryonic stem cell research, arguing that to do so was to make the government complicit in the deaths of the embryos. His detractors bemoaned the limited range of research his policy was to permit and warned that the nation would rapidly fall behind other countries in this important field of scientific research. Over the ensuing years, both sides worked to drive U.S. policy in one direction or another, proposing legislation that would ban all forms of embryonic research or that would permit all forms of therapeutic embryonic stem cell research, if not embryonic research for the purpose of human cloning.

CALIFORNIA HEALTH AND SAFETY CODE §125300–125320 (2002)

The state of California has been at the forefront of efforts to encourage, promote, and support most kinds of stem cell research, including embryonic, adult, and fetal blood stem cell research. On February 15, 2001, state senator Deborah Ortiz (D-Sacramento) filed SB 1272 (later changed to SB 253 for technical reasons) outlining state policy and practice on stem cell research. The bill passed the State Senate on May 2, 2002, by a vote of 21-10; was sent to the State Assembly and passed that body on August 26, 2002, by a vote of 46-27; and was signed into law by Governor Gray Davis on September 22, 2002. The bill amended various parts of the California Health and Safety Code, including paragraphs 123440, 24185, 12115–12117, and 125300–125320, with the last of these sections containing the primary provisions of the act.

Senator Ortiz's bill declared that California's policy on stem cell research was to be that "research involving the derivation and use of human embryonic stem cells, human embryonic germ cells, and human adult stem cells from any source, including somatic cell nuclear transplantation, shall be permitted and that full consideration of the ethical and medical implications of this research be given." To carry out this policy, the bill required that any health care provider involved in providing fertility treatments to supply his or her patients with information about the donation of surplus embryos produced in those treatments. The bill also outlined the specific conditions under which embryos could be donated for the purpose of research and prohibited the sale of embryos or embryonic tissue for research purposes.

STATE OF CALIFORNIA PROPOSITION 71 (2004)

Advocates of embryonic stem cell research in the United States have faced a two-pronged challenge since President George W. Bush laid out federal policy on SCR in August 2001. In the first place, they must ensure that individual state laws allow stem cell research, a situation that often does not exist in many areas of the country. Obtaining legal approval for stem cell research can be achieved by legislative action, by having state legislatures adopt laws that permit stem cell research, or by popular initiative, where such actions are permitted by state constitutions.

Getting stem cell research legalized is only the first step, however. Proponents of SCR must then find a way to obtain the funds needed to support such research. The types of research that are eligible for federal grants are limited by President Bush's 2001 statement, and few states have sufficient funds in their own budget to offer funding for stem cell research.

One of the most popular options for obtaining funding is through bonds that are issued against a state's financial credit, which was what stem cell research supporters in California did in 2004. Proposition 71 called for significant changes (running eight pages in length) to the state constitution and certain laws pertaining to medical research. It called for the establishment of a California Institute for Regenerative Medicine with the charge of regulating stem cell research and providing funding, through grants and loans, for such research and research facilities. The proposition also established a constitutional right for the conduct of stem cell research in the state and specifically prohibited the state's funding of any form of human reproductive cloning research. To carry out the institute's charge, the proposition provided $3 million from the state's general fund to pay for start-up costs and authorized the issuance of general obligation bonds in the amount of $3 billion over a 10-year period, with a maximum expenditure of $350 million in any one year, the bonds to be paid for from the state's general fund.

Proposition 71 was envisioned originally by state senator Deborah Ortiz (D-Sacramento) who had been working for two years to pass legislation in the California legislature to achieve the objectives of what was later to become Proposition 71. She was unsuccessful in her efforts in the legislature and decided to bring the issue to the general public in a proposition to be placed on the November 2004 ballot. The proposition passed handily in that election by a vote of 6,370,852 (59.1 percent) Yes to 4,419,373 (40.9 percent) No. A complete text of Proposition 71 is available online at http://www.voterguide.ss.ca.gov/propositions/prop71text.pdf.

NEW JERSEY PERMANENT STATUTES, TITLE 26, SECTION 2Z-1 (2004)

In recent years, supporters and opponents of stem cell research have become increasingly active at the state level. At least one factor in this trend has been the feeling that President George W. Bush's opposition to SCR is likely to prevent action at the federal level for the foreseeable future, so that the greatest hope for promoting such research may be actions taken within individual states. Thus, grassroots organizations and state legislators in California, Connecticut, Florida, Illinois, Maryland, Massachusetts, New Jersey, New York, Pennsylvania, Texas, and a number of other states have acted to codify approval of and/or governmental support for stem cell research within their own state boundaries. At the same time, organizations and legislators in Alabama, Arizona, Arkansas, Indiana, Iowa, Kansas, Michigan, Missouri, North Dakota, South Dakota, Tennessee, and other states have acted to ban such research and prevent its funding within their states. For example, the state of Nebraska's statute 71-7606 prohibits the use of state funds for the support of human embryonic stem cell research or for any research using human fetal tissue "obtained in connection with the performance of an induced abortion," as well as for "abortion, abortion counseling, referral for abortion, [or] school-based health clinics."

A striking contrast to Nebraska's law is one passed by the New Jersey legislature and signed by Governor James E. McGreevey in 2004. That law declared that "It is the public policy of this State that research involving the derivation and use of human embryonic stem cells, human embryonic germ cells and human adult stem cells, including somatic cell nuclear transplantation, shall: (1) be permitted in this State; (2) be conducted with full consideration for the ethical and medical implications of this research; and (3) be reviewed, in each case, by an institutional review board operating in accordance with applicable federal regulations." The bill specified certain conditions under which stem cell research had to be conducted and made it clear that cloning an embryo for the purpose of human reproduction was unacceptable and was declared to be a crime of the first degree.

The New Jersey act was passed with large majorities in both houses of the state legislature with the strong support of both Governor McGreevey and his successor, Governor Richard J. Codey. The first fruit of the bill's passage was the announcement by Codey in 2004 that the state was establishing the first state-supported and -funded stem cell science research institute in the nation, an institute to be operated jointly by Rutgers University and the University of Medicine and Dentistry of New Jersey. Codey announced that the institute would be funded with an initial grant of $380 million, $150 million from state appropriations and $230 million from a state bond.

STEM CELL RESEARCH ENHANCEMENT ACT OF 2005 (H.R. 810)

Following President George W. Bush's decision in 2001 to limit the range of stem cell research that could be funded by the federal government, legislators from both parties offered a number of bills designed either to (1) confirm and/or extend the president's action or (2) override his action by adopting laws that are more liberal in funding permitted for stem cell research. One of the most successful efforts was House bill 810 (H.R. 810), offered by representatives Mike Castle (R-Del.) and Diana DeGette (D-Colo.) and cosponsored by more than 150 other members of the House of Representatives.

H.R. 810 was introduced by Representative Castle on February 16, 2005. A comparable bill, S. 471, was also introduced into the U.S. Senate by senators Arlen Specter (R-Pa.) and Tom Harkin (D-Iowa). The House bill was reported out of the House Energy and Commerce Committee favorably and was passed by the House on a vote of 238-194 on May 24, 2005. S. 471 was passed by the Senate on a vote of 63-37 on July 18, 2006, but vetoed by President Bush on July 19. The House of Representatives was unable to override the veto.

Both bills took the form most commonly favored by supporters of stem cell research, permitting the funding of such research by the federal government, provided certain conditions are met. Those conditions are that the embryos to be used in research must come from the supply of surplus embryos supplied by in vitro fertilization clinics, but only if donors give their express, informed consent, and that donors not be paid for the embryos. No provision is made for the support of research on embryos that have been specifically produced for experimental use. To make the House's position clear, the bill contains an explicit indication that the surplus embryos can be used "regardless of the date on which the stem cells were derived from a human embryo." This explicit statement presumably was intended to contrast with President Bush's limitation on the use of only those embryos in existence prior to his August 9, 2001, announcement.

The Human Cloning Prohibition Act of 2005 (H.R. 1357)

An example of the type of bills offered by opponents of stem cell research is H.R. 1357, introduced by Representative Dave Weldon (R-Fla.) on March 17, 2005, with 120 cosponsors, after which it was referred to the House Subcommittee on Crime, Terrorism, and Homeland Security. On the same date, a companion bill, S. 658, was introduced in the Senate by Senator Sam Brownback (R-Kans.), with 31 cosponsors. H.R. 1357 prohibited all forms of human cloning, for whatever purpose, including both the reproduction of new humans and the production of embryos for stem cell research. It also prohibited the shipping, receiving, exportation, or importation of any human embryo or any product from a human embryo that has been produced by cloning. No action was ever taken on H.R. 1357.

Stem Cell Therapeutic and Research Act of 2005 P.L. 109-129

The U.S. Congress has also considered a number of bills relating to forms of stem cell research discrete from studies that involve the use of human embryos. Such bills have generally met with greater success than those focusing on embryonic stem cell research. An example is the Stem Cell Therapeutic Research Act of 2005 (H.R. 2520), introduced by Representative Christopher H. Smith (R-N.J.) on May 23, 2005. The bill requires the Secretary of Health and Human Services to "enter into one-time contracts with qualified cord blood stem cell banks to assist in the collection and maintenance of 150,000 units of high-quality human cord blood to be made available for transplantation through the C. W. Bill Young Cell Transplantation Program." (The C. W. Bill Young Cell Transplantation Program is a program named after U.S. Representative C. W. Bill Young (R-Fla.), who was instrumental in creating the U.S. bone marrow donor registry program in 1986. The program is described in section 379 of the Public Service Act, 42 U.S.C. 274l.)

H.R. 2520 passed the House of Representatives by a vote of 431-1 on May 24, 2005, and by the Senate on December 16, 2005. The bill was signed into law by President George W. Bush on December 20, 2005, and has become Public Law 109–129.

General Laws of the Commonwealth of Massachusetts, Chapter 111L (2005)

The state legislature of the Commonwealth of Massachusetts considered the question of stem cell research in 2005 and adopted legislation declaring that the state's policy was to be one that would "actively foster research and therapies in the life sciences and regenerative medicine by permitting research and clinical applications involving the derivation and use of human embryonic stem cells, including research and clinical applications involving somatic cell nuclear transfer, placental and umbilical cord cells and human adult stem cells and other mechanisms to create embryonic stem cells which are consistent with this chapter." The legislation also declared that it was the state's further policy "to prohibit human reproductive cloning."

The Massachusetts legislation took a somewhat tortuous path through both houses of the legislature, beginning as SB 25, sponsored by senate president Robert E. Travaglini (D-First Suffolk and Middlesex), Cynthia Stone Creem (D-First Middlesex and Norfolk), and Harriette L. Chandler (D-First Worcester). That bill was reported out of the senate Committee on Economic Development and Emerging Technologies as SB 2027 and was later retitled SB 2032 (in connection with a comparable bill from the house, bill number 2792), and finally as SB 2039 after a conference committee between the two houses to resolve relatively minor differences in H 2792 and SB 2032.

SB 2039 was passed by both houses of the Massachusetts legislature, by the Senate on April 26, 2005, on a vote of 34-2, and by the House on May 4 on a vote of 119-38. Those votes set up a certain confrontation with Republican Governor Mitt Romney, who opposes stem cell research and had been threatening to veto the legislation. After the legislature defied the governor by passing SB 2039, Romney carried out on his threat, and vetoed the legislation, returning it to the legislature with four recommended changes that he said would make it acceptable to him. (Massachusetts is one of the few states in which governors can follow such a procedure.) The legislature was not convinced by the governor's arguments, however, and both houses voted to override his veto on May 19, 2005, the House by a vote of 112-42, and the Senate by a vote of 34-2. Since the legislation had been given "emergency" status, it became law immediately upon the legislature's vote.

COURT DECISIONS

As noted above, the field of stem cell research is still very young, and little case law exists on the field. The most useful court cases currently available for citation are those that have dealt with the rights of fertilized eggs, "pre-embryos," embryos, and fetuses in other contexts, primarily with regard to abortion (in which such an entity is destroyed) and in vitro fertilization (in which a dispute exists over the ownership and ultimate disposition of frozen "pre-embryos").

ROE V. WADE (1973), 410 U.S. 113

Background

One of the most famous legal cases in American history, *Roe v. Wade*, was heard by the U.S. Supreme Court on December 13, 1971, and October 11, 1972, after an anonymous female identified only as "Jane Roe" brought a class-action suit against the state of Texas, arguing that its criminal abortion laws were unconstitutional. Those laws prohibited the performance of an abortion for any reason whatsoever except the saving of a pregnant woman's life. A three-judge district court agreed with the plaintiff and found the Texas laws void as "vague and overbroadly infringing those plaintiffs' Ninth and Fourteenth Amendment rights." That court's decision was appealed directly to the U.S. Supreme Court, bringing to a head a long-simmering and fundamental debate in American society as to the legality of abortions.

Legal Issues

In its deliberations, the Court first had to contend with certain technical issues involving two other cases with which the original Texas case had been bracketed. The first involved a couple known only as "the Does" who, although not pregnant, were challenging the state of Georgia's laws against abortions. The second concerned a Texas attorney named James Hallford who had already been convicted twice of performing illegal abortions, and also challenged the state of Texas's antiabortion laws. The court decided that neither the Does nor Hallford had standing in this case, and dismissed their claims from further consideration.

The fundamental legal issue that remained, then, was simple in concept: Does a woman have a constitutional right to abort a child that she is carrying? In fact, the U.S. Constitution does not explicitly address the issue of abortion. But lawyers for Roe in this case argued that the right to abortion was inherent in a number of constitutional amendments, including the First, Fourth, Fifth, Ninth, and Fourteenth Amendments. The key amendments among these were the Fourteenth Amendment, the so-called due process amendment, and the Ninth Amendment, which reserves to the people all rights not specifically granted to the federal government in other parts of the Constitution. Roe's attorneys argued that these two amendments guaranteed a right of privacy to citizens that, other constitutional restrictions notwithstanding, included the right to abortion.

The issue was complicated by the question as to whether, when, or under what circumstances the embryo, fetus, or unborn child being carried by a pregnant woman itself became a "person," with all of the constitutional rights then due it. At that point, the constitutional rights of the pregnant woman alone could no longer be considered by themselves; they would have to be weighed against the constitutional rights of the new "person" she was carrying.

Decision

In reaching its decision in *Roe v. Wade*, the Court reviewed in great detail ancient attitudes toward abortion; the Hippocratic Oath; common law, English statutory law, and American law on abortion; and the views of the American Medical Association, the American Public Health Association, and the American Bar Association. Based on its reading of these records, the Court made two fundamental decisions about the constitutional status of abortion in the United States. First, the Court decided that the Constitution did, in fact, confer a "right of privacy" to individuals and included within that right was the right for a woman to do with her own body as she chose to do, i.e., abort a child that she was carrying. Specifically, it said that "[t]his right of privacy, whether it be founded in the Fourteenth Amendment's concept of personal liberty and restrictions upon state action, as we feel it is, or, as the District Court determined, in the Ninth Amendment's reservation of rights to the people, is broad enough to encompass a woman's decision whether or not to terminate her pregnancy."

Second, the Court resolved the issue as to the point at which an embryo/fetus becomes a person by separating pregnancy into three distinct trimesters, each representing a more advanced stage of development of the unborn child. It concluded that the chances of an embryo's/fetus's surviving outside the uterus during the first trimester was so small that it could not, at that point, legally be considered to be a "person" and that it had, therefore, no constitutional rights. Under the circumstances, a state could pass no laws prohibiting abortion during that first trimester.

By contrast, evidence suggested that a fetus might be able to survive during the second trimester and, hence, states could pass laws banning abortions in the interest of protecting a mother's health or life. And, given the substantial likelihood that a third-trimester child could survive outside the uterus, states were allowed to pass any type of legislation restricting or banning abortion during this period of pregnancy.

Impact

The most important impact of *Roe v. Wade* was, of course, in the area of abortion itself. All state laws banning abortion during the first trimester were immediately made invalid, and most laws placing restrictions on second-trimester abortions were also brought into serious question. The Court's decision did not, of course, end the debate over abortion, which continues to this day with at least as much intensity as it did in the early 1970s.

With respect to the issue of stem cell research, the Court's decision did not resolve the question as to when life begins and, hence, the legal status of embryos. It adopted a philosophy that the process by which a fertilized cell becomes a human being, with all of its constitutional rights in the United States, is an ongoing process in which the fertilized egg first becomes an embryo, then a fetus, then a human being. In this regard, *Roe v. Wade* has had only a modest impact on the current debate over the legality and morality of stem cell research.

York v. Jones (1989), 717 F. Supp. 421 (E.D. Va. 1989)

Background

Steven York and Risa Adler-York were married in 1983 and, for two years, attempted to become pregnant, using natural coital procedures, but without success. In 1986, the couple consulted with Dr. Howard W. Jones and his associates at the Howard and Georgeanne Jones Institute for Reproductive Medicine in Norfolk, Virginia, about the possibility of using in vitro fertilization (IVF) techniques to achieve pregnancy. Prior to beginning the IVF procedure, the Yorks signed a number of consent forms that are commonly part of the IVF agreement. Included among the consent forms was an agreement specifying the disposition of any unused embryos produced by the IVF procedure that were cryopreserved and stored.

The IVF procedure used at the Jones Institute eventually resulted in the production of six fertilized eggs, five of which were implanted in Mrs. York's uterus. The sixth was frozen and stored at the Jones Institute. In May 1988, the Yorks, who by this time had moved to California, contacted the Jones Institute and requested that the single remaining frozen fertilized egg (which the court called a "pre-zygote," an incorrect term) be transferred to the Institute for Reproductive Research at the Good Samaritan Hospital in Los Angeles. The Jones Institute refused to do so, claiming that it retained rights to the fertilized egg through the consent forms signed by the Yorks.

At this point, the Yorks filed suit in the U.S. District Court for the Eastern District of Virginia, claiming a breach of contract with the Jones Institute. The case was decided on July 10, 1989.

Legal Issues

York v. Jones was one of the earliest cases in which courts have been faced with the issue as to how a very early stage embryo is to be regarded in the law. The case was couched in terms of a property rights issue, with the plaintiffs arguing that they owned the single remaining fertilized egg stored at the Jones Institute, as they might own any other piece of property and that, therefore, they had the right to transfer that fertilized egg to their new home in California, as they would any other piece of property. In response, the Jones Institute claimed that the consent forms signed by the Yorks did not specifically grant them the right to take position of any unused fertilized eggs and that, therefore, the institute had the right to retain any such eggs and to use them as they liked.

Decision

The court ruled that plaintiffs' interpretation of the consent agreements they signed with the Jones Institute was, in fact, correct, and that the remaining frozen fertilized egg was properly construed as their property and could be transferred to California as could be any other item of property they owned. In another portion of the case, the court also ruled that the Jones Institute was not immune from litigation under the Eleventh Amendment to the U.S. Constitution, as defendants had claimed, because the institute is "not an arm of the Commonwealth [of Virginia]" and has been granted "a high degree of autonomy" by the commonwealth.

Impact

York v. Jones provided a useful test case at a time when courts were just beginning to attempt to answer questions regarding the legal status of the embryo and the fetus. Most later cases, including those discussed below, made reference to the case and cited its views on the property rights of donors over the fertilized eggs they produced.

DAVIS V. DAVIS (1992), SUPREME COURT OF TENNESSEE, 842 S.W.2D 588, 597

Background

The two individuals in this case, Junior Lewis Davis and Mary Sue Davis, met while both were in the U.S. Army in Germany in 1979 and were married when they returned to the United States a year later. After an extended period of time during which they unsuccessfully attempted to become pregnant, the couple decided to seek advice and treatment from an in vitro fertilization (IVF) company. IVF treatment resulted in the production of eight fertilized eggs, the first of which was implanted into Mrs. Davis's uterus, but without a successful pregnancy resulting. Before additional implantations could be attempted, Mr. Davis filed for divorce. The couple successfully resolved all issues relating to disposition of property during the divorce hearings with the exception of the seven remaining frozen embryos. Mrs. Davis asked that the embryos remain frozen in the hopes that future implantations might result in a pregnancy, while Mr. Davis asked that the embryos be destroyed, not wishing to have a successful pregnancy result when he could no longer act as a child's father.

The trial court found for Mrs. Davis, ruling that the embryos were human beings from the moment of fertilization and that she had the right to raise them as her own children. Mr. Davis appealed that ruling, and the Court of Appeals reversed the trial court's decision, finding that Mr. Davis had a "constitutionally protected right not to beget a child where no pregnancy has taken place" and that the state had no compelling interest in ordering that the embryos be implanted into Mrs. Davis's uterus without Mr. Davis's consent. Finally, the case was taken by the Tennessee Supreme Court, which eventually affirmed the Court of Appeals decision.

Legal Issues

The state supreme court's deliberations in this case were made more difficult for three reasons. First, the Davises had signed no agreement or letter of intent with the IVF company at the time they began treatment specifying the disposition of any unusued ("surplus") embryos. Second, the state of Tennessee had not yet adopted legislation specifying the disposition of surplus embryos that were no longer wanted by the donors. Third, and perhaps most important, no case law existed to provide the court with precedents on which to base its deliberations and its decision: *Davis v. Davis* was, in fact, the first case in the United States in which a state supreme court had to decide the disposition of surplus frozen embryos.

In its decision, the court pointed out that perhaps the most critical factor involved in its deliberations was how the seven four- and eight-celled entities were to be defined, both scientifically and legally. After a lengthy testimony by a number of scientific experts on embryology and based on its own analysis of that testimony, the court chose to use the term *pre-embryos* for the four- and eight-celled entities. In doing so, it formed the basis for the second legal decision with which it was faced: are these four- and eight-celled entities persons or are they property? The resolution of that question was critical because if the pre-embryos were to be regarded as persons, as they were by the original trial judge, they deserved the protection of the law and the appellant (Mrs. Davis) would prevail. If they were not to be regarded as persons, and could, therefore, be classified as property, then the appellee (Mr. Davis) might prevail.

Decision

The state supreme court concluded that the four- and eight-cell entities, the pre-embryos, at dispute in this contest did not meet the standards of "personhood" set either by Tennessee or federal law. Instead, they occupied some intermediary category between "persons" and "tissue" that deserved greater respect than the latter, but less respect than the former. In such a case, the dispute before the court had to be resolved in terms of which person, Mr. Davis or Mrs. Davis, had the greater claim on the pre-embryos. Since Mrs. Davis had concluded that she no longer wanted or planned to have the pre-embryos implanted in her own uterus, but wanted to donate them to another couple, her claim was inferior to that of her former husband, the court said, and his claim was to be preferred. The court affirmed the appeals court's decision in the case and ordered the Knoxville Fertility Clinic "to follow its normal procedure in dealing with unused preembryos."

Impact

As *Davis v. Davis* worked its way through the Tennessee courts, the case drew more and more public interest, with supporters of both sides becoming increasingly agitated and vociferous. Indeed, after the court's decision was announced, judge Dale Young, who wrote the court's opinion, felt compelled to add an appendix, which he entitled "Some Fundamental Principles Utilized by the Court" to explain to the general public the basis for the court's decision. He acknowledged the anguish that the case had caused for both appellee and appellant and their supporters, but made it very clear that the court was obligated to make its decision strictly on the basis of the facts presented to it and on the existing law. "Cases are to be decided by the head," he concluded, "not by the heart." Young's appendix itself has become a classic statement for students of the law as to the fundamental principles on which any court decision must be based.

In addition, as the first decision of its kind in the United States, *Davis v. Davis* has obviously had an influence on the thinking of other state courts in determining their actions on similar problems of the disposition of pre-embryos and embryos, of frozen sperm, and of similar prebirth materials. Such courts commonly mention *Davis v. Davis* as a precedent in their own decisions on such matters. But the case has had an important impact for another reason also. The Tennessee Supreme Court concluded that individuals have a constitutional right to procreate and *not* to procreate. When these two rights are in conflict, as they were in the case of Mr. and Mrs. Davis, then the court must weigh the relative harm caused to each person's right of procreation or nonprocreation and determine which individual will suffer the least harm by its decision.

Finally, *Davis v. Davis* is of considerable interest in case law dealing with stem cell research because it is essentially the only major court case in which a court refers specifically to note that pre-embryos or embryos have a "potential for life," and may, therefore, possibly be viewed as something other than purely a matter of property.

SANTANA V. ZILOG (1996), 95 F.3D 780, 783 (9TH CIR.)

Background

Jodene M. Santana and Michael Santana, husband and wife, brought a wrongful death complaint in 1994 against a company for which Mrs. Santana worked, Zilog, Inc., claiming that her working conditions had been responsible for six miscarriages she had experienced between 1988 and 1993. Mrs. Santana worked in a facility at Zilog's Nampa, Idaho, computer assembly plant in a room called "The Fab," where a number of potentially dangerous chemicals were in routine use. After Mrs. Santana was transferred in December 1993 to a work space outside The Fab where such chemicals were not in use, she successfully delivered a child with which she had been pregnant at the time of her transfer. She had also delivered three other children prior to being employed at Zilog. The Santanas complained in their suit that exposure to chemicals used in The Fab had been responsible for Mrs. Santana's six miscarriages. The Santana's suit having been filed, Zilog filed a motion with the district court to have the case dismissed on the grounds that the fetuses miscarried by Mrs. Santana were, in fact, nonviable fetuses, incapable of surviving outside her body and not, therefore, "persons" in a legal context. The district court agreed with Zilog's reasoning and dismissed the case. The Santanas appealed that decision to the U.S. 9th District Court of Appeals. The case was heard on February 5, 1996, and decided on September 4, 1996.

Legal Issues

The basic question faced by the court in this case is whether the six miscarriages experienced by Mrs. Santana satisfied the conditions of the state of Idaho's wrongful death statute. That statute says, in part, that "[w]hen the death of a person is caused by the wrongful act or neglect of another, his or her heirs or personal representatives on their behalf may maintain an action for damages against the person causing the death. . ." The issue to be determined, then, is whether the six aborted fetuses could be defined as "persons." If they could, then the wrongful death statute was applicable. If they could not, the statute did not apply, and the Santanas had no case.

Decision

In considering its decision, the court of appeals relied on an extensive review of all existing case law dealing with prenatal injury to fetuses that were eventually born alive. It found that in the vast majority of such cases, courts granted relief to the injured party. When the fetus to whom injury had been caused was *not* born alive, however, the results were very different. In such cases, courts, almost without exception, failed to recognize the fetus as a "person" and said that wrongful death and injury statutes did not apply. The supreme court of West Virginia was the only upper court in the nation that had granted a nonviable fetus the right of "personhood" without specific direction from the state legislature to do so.

The key criterion in cases such as the one before it, the court of appeals decided, was whether or not the fetus was "capable of sustaining an independent, separate existence from its mother." It adopted a previously stated view that "A nonviable fetus is not a distinct entity; rather, its life is an integral part of its mother's life" (cited in Humes, 792 P.2d at 1037). Based on the evidence available to it, then, the court determined that the most likely point at which viability occurs is "at about week 23 to 24 of gestation." Since the longest of the six terminated pregnancies was only 17 weeks, the court decided that none of the aborted fetuses was entitled to the legal status of "personhood," the complaint was improper, and the lower court's decision was upheld.

Impact

One of the core ethical and legal issues related to embryonic stem cell research is the status of the embryo. Although *Santana v. Zilog* does not address that issue directly, it serves part of the existing legal framework in which the issue is currently interpreted. Courts at all levels in all parts of the nation have, for the most part, been unwilling to grant the status of "personhood" on fetuses that have not yet reached the age of viability, an age of "about 23 to 24 weeks after gestation." Under such circumstances, the embryo, which is clearly much younger and less viable than the fetus, has very seldom been regarded by the courts as a person. Instead, as the following cases demonstrate, the embryo is far more likely to be viewed as property, whose fate is to be determined by property law, rather than law dealing with the civil rights of a person.

KASS V. KASS (1998), 696 N.E.2D 174

Background

Maureen and Steven Kass were married in 1988, and immediately began to try becoming pregnant. When, after five years of such efforts, they were unsuccessful in becoming pregnant, the Kasses enrolled in the in vitro fertilization (IVF) program at John T. Mather Memorial Hospital in Port Jefferson, Long Island. As part of that program, the Kasses were required to sign four consent forms dealing with various aspects of the IVF procedure they were about to begin. One of the consent forms had to do with the disposition of any fertilized eggs that were not actually used in the IVF program in the event they no longer wished to initiate a pregnancy or were unable to decide what to do with the stored, frozen fertilized eggs.

About three weeks after initiating the IVF program, the Kasses began a divorce procedure. As part of that procedure, they jointly signed a memorandum indicating that they wished the fertilized eggs not yet implanted in Mrs. Kass's uterus to be disposed of as prescribed in the original consent form. After another three weeks, Mrs. Kass changed her mind and asked that the fertilized eggs not be destroyed, but that they be saved for possible future implantation. Her husband opposed this decision, and when the issue was still not resolved at the time of the couple's divorce in May 1994, Mrs. Kass filed the lawsuit described here.

Legal Issues

In analyzing the legal issues involved in this case, the court pointed out that the process of in vitro fertilization was still very young, and that few state and no federal laws existed covering IVF-related issues. New York, where the case was being heard, was one of those states with no laws controlling the disposition of unused ("surplus") fertilized eggs. The court also noted the paucity of case law, citing *Davis v. Davis* as one of the few precedents on which it could base its deliberations.

In contrast to the lack of guidance from law and court cases, the court noted that a relatively large body of commentary existed on the issues before it. The principles involved in this body of commentary could, the court said, be divided into four general categories: (1) suggestions that ownership of the fertilized eggs should be allocated to one or the other of the disputants, depending on the specific circumstances of a case; (2) acceptance of an implied contract between two people to procreate as evidenced by their willingness to participate in an IVF program, a contract that can then be treated under existing law; (3) recognition that the progenitors (producers of the fertilized egg) hold a "bundle of rights" that can be exercised through a joint disposition agreement; and (4) articulation of a "default rule," that says that, even with a disposition agreement in place, no embryo should be implanted, destroyed, or used in research over the objection of one of the progenitors. The court's ultimate decision in this case, then, represented a review of these four legal philosophies about spare embryos, with its own decision as to which was (or were) most relevant to the specific facts in the case.

Decision

The court decided in this case to view the dispute between the Kasses as essentially a contract dispute. It pointed out how very important it is for a couple undertaking in vitro fertilization to "think through possible contingencies and carefully specify their wishes in writing." The court recognized that no such agreement can possibly foresee all possible contingencies that might arise in the future. But, for that very reason, courts have a special obligation not to ignore the clear arrangements that couples have made in IVF agreements and to follow the exact letter of those agreements. Emphasizing the "contractual" interpretation of the case, the court pointed out that "[t]he subject of this dispute may be novel but the common law principles governing contract interpretation are not." After deciding that Mrs. Kass's right to privacy was not being violated by *Roe v. Wade* consideration, the court ruled that the original agreement regarding the ultimate fate of the fertilized eggs had to be honored, and it affirmed the lower court's decision that the Kasses's fertilized eggs could be released for donation to the original IVF program for "approved research purposes." It summarized its position by saying that "agreements between progenitors, or gamete donors, regarding disposition of their pre-zygotes should generally be presumed valid and binding, and enforced in any dispute between them."

Impact

As one of the earliest court decisions dealing with in vitro fertilization and the status of the embryo, *Kass v. Kass* has obviously been considered and cited by later court cases faced with similar questions. In later cases, some courts have agreed with the New York court's interpretation of the issues and its decision, while other courts have acknowledged the New York argument, but then chosen to follow a different line of reasoning in deciding the cases before them. *A.Z. v. B.Z.* (discussed later) is an example of such an instance in which a court recognized the contractual nature of the dispute between the litigants, but chose to ignore the contract for other reasons.

Kass v. Kass has also been used by later courts in making decisions far removed from in vitro fertilization or the status of the embryo. For example, in a case heard before the New York Supreme Court in 2000 (*Callahan v. Carey*, 42582/79, Sup Ct NY County (New York 1979)), the court cited the lower New York court's reasoning on contractual obligations in a case dealing with an agreement for housing homeless people by the state of New York, an agreement on which it later tried to reneg.

A.Z. v. B.Z., 431 Mass. 150, 725 N.E.2d 1051 (2000)

Background

A. Z. and B. Z. (initials were used to protect anonymity) were a married couple who met and were married in 1977 while both served in the U.S. Army and were stationed in Virginia. In their efforts to become pregnant, B. Z. developed an ectopic pregnancy that resulted in her having to have surgery, during which her left fallopian tube was removed. The couple subsequently decided to pursue in vitro fertilization (IVF) as a way of becoming pregnant, and they participated in an IVF program from 1988 to 1991 that resulted in the birth of twins in 1992. At the time, two vials of frozen fertilized eggs, which the court later designated as *pre-embryos*, remained frozen in storage.

Three years later, the wife decided that she wanted another child and asked that one of the remaining vials be thawed and a pre-embryo implanted in her uterus. She took this action without notifying her husband, who learned of the procedure only when he received insurance forms relating to the implantation. During this period, the relationship between husband and wife was deteriorating and divorce proceedings were initiated by A. Z. in August 1995. As part of that procedure, he filed a motion requesting a restraining order preventing his wife from using any more of the remaining pre-embryos. The case eventually worked its way through the Massachusetts court system and was heard by the Supreme Judicial Court of Massachusetts on November 4, 1999, and decided on March 31, 2000.

As is generally the case, A. Z. and B. Z. were required to sign a number of consent forms at various stages of the IVF treatment. Some of those forms included instructions for the disposition of the fertilized eggs in case of certain future contingencies. On all of these forms, the couple chose, in case of divorce or separation, to have the pre-embryos returned to the wife for possible future implantation. Some legal issues arose regarding the validity of these forms, but those issues turned out not to be germane to the court's final decision in the matter.

Legal Issues

In its decision, the court pointed out that the field of in vitro fertilization was still in its infancy and that only three states (Florida, New Hampshire, and Louisiana) had laws dealing specifically with IVF and the rights of donors and fertilized eggs (by whatever name they might be called). The court also cited two earlier cases, *Davis v. Davis* and *Kass v. Kass*, as possible precedents for its own deliberations. It also noted that the present case was apparently the first recorded instance "in which a consent form signed between the donors on the one hand and the clinic on the other provided that, on the donors' separation, the preembryos were to be given to one of the donors for implantation." The case, therefore, raised new issues as to how the frozen pre-embryos were to be disposed of.

Decision

In its decision, the court ruled that the original consent agreement between A. Z. and B. Z. on the one hand and the IVF clinic on the other was unenforceable. They listed a number of reasons for this decision. First, they said that the forms signed by the donors were intended to provide information primarily on the risks and benefits of IVF technology, and that the disposition conditions were secondary to that purpose. Second, it pointed out that the forms had no duration provision, thus ignoring the possibility that conditions might change significantly over a period of years, making the original agreement no longer tenable or reasonable. Third, it said that phrases in the original agreement mentioning "separation" of the couple was unclear and could mean anything from voluntary physical separation to divorce.

Fourth, the court indicated that the way in which the forms were signed (on some occasions, the husband signed a blank form) made the agreement unenforceable. Finally, it ruled that the forms themselves were not legal documents that could dictate the fate of the pre-embryos in case of some other relevant legal action, such as the couple's divorce. These points having been considered, the court ordered a permanent injunction against the wife's having access to or using the frozen pre-embryos without her ex-husband's specific permission.

Impact

One of the most significant sections of the court's decision was its discussion of the effect of allowing B. Z. to have the couple's fertilized eggs implanted into her uterus, with the subsequent possibility of her later becoming pregnant. Such a scenario would mean that A. Z. would, in effect, be forced into becoming a father, when he had expressly indicated his unwillingness to do so. At the conclusion of its decision, the court said that "In this case, we are asked to decide whether the law of the Commonwealth may compel an individual to become a parent over his or her contemporaneous objection. The husband signed this consent form in 1991. Enforcing the form against him would require him to become a parent over his present objection to such an undertaking. We decline to do so."

The court's decision thus broke ground in one regard, and confirmed a growing trend in a second. In the former case, the court molded the notion of "enforced parenthood" when one person (A. Z. in this case) no longer wished to become a parent, even though the other member of the couple (B. Z.) did. "[W]e would not enforce an agreement," the court said, "that would compel one donor to become a parent against his or her will." The trend to which the court subscribed in its decision was the view of the pre-embryo essentially as property, whose fate could be decided by standard contract law. Like the Tennessee and New York court cases discussed above, the Massachusetts court declined to consider the pre-embryo as a "person" with any rights of its own in a dispute over its ultimate fate.

As might be expected, the court's decision added fuel to the fire of controversy over the status of the fertilized egg. Some commentators on that decision pointed out that it contradicted the beliefs and teachings of a substantial number of the American population, who believed that human life begins at the moment of conception not, as the court's decision implies, at the moment of implantation. As one writer observed, "The most disturbing aspect of the opinion is the value that it assigns to the children involved—that is, none at all in the law's eyes."[4] Within the context of the debate over stem cell research, the court's decision tends to diminish the view that the use of four- and eight-celled entities (the so-called pre-embryos in this decision) are not to be regarded as forms of human life and are not worthy of protection by civil and criminal law.

---

[1] Dianne N. Irving, "What Is 'Bioethics'? (Quid est 'Bioethics'?)", in *Life and Learning X: Proceedings of the Tenth University Faculty for Life Conference*, ed. Joseph W. Koterski (Washington, D.C.: University Faculty for Life, 2002), pp. 1–84.

[2] Quoted in Randy Engel, "The Case of the Missing Moratorium: A Tale of Criminal Politics and Murderous Science." The Catholic Resource Network. Available online. URL: http://www.ewtn.com/library/PROLIFE/MISSMORA.TXT. Downloaded on June 15, 2005.

[3] "The Balanced Budget Downpayment Act, I." Public Law 104-99. Available online. URL: http://frwebgate.access.gpo.gov/cgi-bin/getdoc.cgi?dbname=104_cong_public_laws&docid=f:publ99.104. Downloaded on August 8, 2005.

[4] Daniel Avila, "Massachusetts Court Rules in Frozen Embryo Case," *National Right to Life News*, April 2000. Available online. URL: http://www.nrlc.org/news/2000/NRL04/mass.html.

CHAPTER 3

CHRONOLOGY

This chapter presents a chronology of major events in the history of stem cell research, including scientific and technological advances, as well as political, social, ethical, and other events related to research in this field.

1665

- English physicist Robert Hooke examines the structure of a piece of cork through a microscope and observes tiny pockets that he calls cells or pores, the first recognition of these structures as the basic unit of life.

1712

- French physicist René Réaumur publishes his classic work on the ability of crawfish and crabs to regenerate claws and limbs in the *Mémoires de l'Académie Royale des Sciences.*

1744

- Abraham Trembley, Swiss scientist and philosopher, publishes a report of his extensive studies on the regeneration of hydra in his book *Mémoires pour servir à l'histoire d'un genre de polypes d'eau douce à bras en forme de cornes* (Memoirs concerning the natural history of a type of freshwater polyp with arms shaped like horns).

1827

- German naturalist Karl Ernst von Baer discovers that mammalian life originates with the fertilization of an egg cell.

1838

- German botanist Matthias Jakob Schleiden suggests that the basic structural unit of all plants is the cell.

Chronology

1839

- German physiologist Theodor Schwann hypothesizes that cells are the basic structural unit of all animals. This hypothesis, along with Schleiden's similar suggestion for plants in 1838, constitutes the basis of the cell theory.

1855

- German pathologist Rudolf Virchow proposes a fundamental law of cell behavior, *omnis cellula a cellula*, "all cells arise from cells."

1869

- Pope Pius IX declares that the process of ensoulment begins at conception, establishing Roman Catholic doctrine that human life begins at that moment.

1878

- Austrian physician and researcher S. L. Schenk attempts to fertilize a human egg outside the human body in an attempt to assist a woman unable to have children by normal coitus. The experiment fails, but it is the first recorded effort to achieve ex utero fertilization.

1888

- German embryologist Wilhelm Roux carries out an experiment in which he destroys one of the two cells in a young frog embryo. The remaining cell develops into a deformed frog, convincing Roux that each of the two original cells contains only half of the frog's complete genetic map. Roux's results are later contradicted by experiments conducted by Hans Driesch.

1891

- English biologist Walter Heape performs the first known case of embryo transfer when he flushes preimplanted embryos from the oviducts of one female rabbit and transplants them into the oviducts of a second female rabbit. The second rabbit produces a mixed litter of two angora and four Belgian rabbits.

1892

- German embryologist Hans Driesch separates individual embryonic cells from two- and four-cell sea urchin embryos. He finds that each of the individual cells develops normally into a complete adult, contradicting the results of Roux's experiments conducted only four years earlier.

1902

- Austrian botanist Gottlieb Haberlandt proposes the theory of totipotency for plant cells, namely, that every cell in a mature plant has the capability to change back into an embryonic form that has the potential to then grow and differentiate into every type of cell from which the mature plant is made.

1903

- Herbert J. Webber, at the U.S. Department of Agriculture, invents the term *clon* to describe "a colony of organisms derived asexually from a single progenitor." Other biologists soon adopt the term and change its spelling to its modern form of clone.

1909

- Russian physician and biologist Alexander A. Maximow hypothesizes the existence of immature cells within bone marrow—he calls them *gemeinsame Stamzellen*, or *common stem cells*—with the capability of generating new blood cells. Maximow's hypothesis is largely rejected and ignored for nearly 50 years.

1928

- Hans Spemann, German embryologist, performs the first somatic cell nuclear transfer experiment, transferring the nucleus from one salamander egg into the enucleated egg cell of a second salamander. For this research, Spemann was later awarded the 1935 Nobel Prize in Physiology or Medicine.

1953

- Robert Briggs and Thomas J. King, American embryologists, remove the nuclei from frog embryos and insert them into unfertilized frog eggs from which the nuclei have been removed. They find that the eggs then begin to develop normally, growing through early embryonic stages, although seldom hatching into tadpoles. They further demonstrate that the effectiveness of the technique is a function of the age of the embryos used, with nuclei taken from younger embryo able to produce healthier eggs that develop more normally than with nuclei taken from older embryos.

Chronology

1956

- E. Donnall Thomas leads a surgical team that performs the first successful bone marrow transplant on a human. The patient achieves full remission after receiving a transplant from an identical twin.

1958

- Cornell University botanist Frederick C. Steward demonstrates that it is possible to grow a complete new carrot plant starting with a single somatic cell removed from a mature carrot plant. The experiment shows the ability of a mature, completely differentiated somatic plant cell to revert to a similar, totipotent form that is then able to grow through the complete cycle of redifferentiation and development.

1961

- Daniele Petrucci, an Italian embryologist, conducts one of the first successful in vitro fertilization experiments. He fertilizes an egg on a laboratory dish and embeds it in an "artificial womb." The fertilized egg appears to develop normally for 29 days, at the end of which Petrucci claims to have detected a heartbeat. He then decides to destroy the embryo because of profound ethical questions raised by the research and because the embryo itself had become enlarged, deformed, and, according to Petrucci, "a monstrosity."

1962

- English biologist John B. Gurdon performs the first successful cloning experiments with animals by removing the nuclei of cells taken from stomach tissue of a tadpole and transplanting them into the enucleated eggs of frogs. Some small number of eggs eventually develop into young tadpoles, although they do not survive for more than a few days.
- American researcher Joseph Altman begins publishing a series of papers suggesting that neurogenesis (the formation of new nerve cells) occurs in the brain. Prior to this time—and, for many scientists, even following the publication of Altman's papers—the prevailing wisdom among neuroscientists was that an organism is born with all the brain cells (neurons) it will ever have and that no new neurons are ever produced in the brain.

1963

- Three Canadian researchers, Andrew Becker, Ernest McCulloch, and James Till, report on the existence of a group of cells in mouse spleens with the ability to regenerate rapidly and evolve into blood cells, the first direct evidence of the existence of hematopoietic blood cells in a mammal.

1964

- Two American medical researchers, G. Barry Pierce and Lewis Kleinsmith, demonstrate that teratomas are caused by a specific type of stem cell, known as an embryonal cancer cell.

1968

- British physiologists Robert Edwards and Barry Bavister successfully fertilize a human egg with human sperm in a petri dish. The procedure provides the fundamental technology later used for in vitro fertilization.

1973

- ***January 22:*** The U.S. Supreme Court announces its decision in the case of *Roe v. Wade*, setting the nation's policy on abortion that is essentially intact today. In that decision, the Court avoids a specific definition as to when human life begins but allows unfettered access to abortions by women during the first trimester of pregnancy. It permits some forms of restrictions on abortions during the second trimester, and even more restrictions by states during the third trimester, a pattern reflecting the Court's view that the origin of human life is an evolutionary process, rather than an event that can be specified at a particular moment in an organism's history.

1974

- ***July 12:*** Congress passes the National Research Act, setting out the basic legal and ethical standards to be used in biomedical and behavioral research projects with human subjects. The act created the National Commission for the Protection of Human Subjects of Biomedical and Behavioral Research to oversee human experimentation in the United States and established a moratorium on the use of embryos and fetuses and embryonic and fetal tissue.

1975

- American biologists Howard Green and James Rheinwald develop a method for growing cells on a culture called 3T3 made of irradiated mouse fibroblasts. The Green-Rheinwald method later becomes the standard procedure for maintaining stem cells in a proliferative, nondifferentiating state in vitro over long periods of time.

Chronology

1977

- ***September:*** Secretary of Health, Education, and Welfare (HEW) Joseph A. Califano appoints an Ethics Advisory Board (EAB) to provide the department with ethical advice about a number of scientific, medical, and health issues involving humans.

1978

- ***July 25:*** Baby Louise Brown is born in England, the first child conceived by in vitro artificial insemination. She becomes popularly known as the world's first "test-tube" baby.

1979

- ***April 18:*** The National Commission for the Protection of Human Subjects of Biomedical and Behavioral Research, authorized under the National Research Act of 1974, issues a report on the general principles and guidelines to be followed in research on human subjects. The report, commonly called the Belmont Report, is important because it is the first official document in U.S. history discussing the legal status of the embryo and its treatment in scientific research.
- ***May 4:*** The Ethics Advisory Board issues its report on in vitro fertilization for the purposes of producing human embryos. The board recommends that federal funds be permitted for research involving embryos, provided that research takes during the first 14 days of embryonic development, and that all gamete donors are married couples.

1980

- The original charter for the Ethics Advisory Board expires and is not renewed. Lacking any agency for approving the federal funding of research on embryos in the United States (the EAB's original responsibility), a de facto moratorium on such research exists until 1993.

1981

- Martin Evans and Matthew Kaufman in the United Kingdom and Gail Martin in the United States report the isolation and culturing of murine embryonic stem cells (stem cells obtained from mice), the first case in which embryonic stem cells are obtained from an animal.

1982

- American ornithologists Steven Goldman and Fernando Nottebohm demonstrate the existence of neuronal stem cells in the brain of canaries that are responsible for the constant process of neurogenesis (the formation of new nerve cells). The experiment is particularly interesting because of the long-held belief that the brain is not capable of growing new nerve cells.

1984

- The Committee of Inquiry into Human Fertilisation and Embryology, commonly known as the Warnock Committee, issues its report on infertility research and services in the United Kingdom. The report deals with a wide range of topics, including artificial insemination; sperm, ovum, and embryo donation; in vitro fertilization; use of frozen sperm, ova, and embryos; research with embryos; informed consent of participants; disposal of embryos; National Health Service planning; legal implications; and surrogate motherhood arrangements.

1987

- A research team at the Walter and Eliza Hall Institute of Medical Research of the Royal Melbourne Hospital, in Australia, under the direction of David Paul Gearing, announces that they have been able to maintain murine embryonic stem cells in an undifferentiated state using a feeder layer consisting of leukemia inhibitory factor (LIF). The discovery is important because prior to this time, essentially the only effective feeder layer available was made of irradiated murine embryonic fibroblast cells.

1988

- The first cord blood transplant is used to treat a five-year-old Parisian boy with Fanconi's anemia, a genetic disorder in which a person's bone marrow fails to produce adequate numbers of all types of blood cells. Cord blood is blood obtained from the umbilical cord of a newborn baby. The boy received a cord blood transplant from his sister and is still alive with no symptoms of the original disorder.
- Irving Weissman and his colleagues at Stanford University develop a system for recognizing hematopoietic stem cells in mice, providing a reliable method for isolating such cells from other cells present in bone marrow. Four years later, Weissman's team creates a similar system for the identification of human hematopoietic stem cells.
- ***December:*** A committee of the National Institutes of Health (NIH) votes 19-2 to renew government funding of research on fetal tissue transplantation. The recommendation is never put into effect.

Chronology

1990

- The British Parliament passes the Human Fertilisation and Embryology Act permitting research on donated embryos for certain limited purposes, including studies on infertility and the detection of birth defects.

1993

- The Canadian Royal Commission of New Reproductive Technologies, which had begun its work in 1983, issues a 14-volume report that includes 293 recommendations for national policy and procedures regarding the use of a variety of artificial reproductive technologies. For a variety of political reasons, the government does not begin to act on the most important of these provisions for over a decade.
- The first cord blood transplant between unrelated adults with acute lymphoblastic leukemia (ALL) is conducted by Dr. Joanne Kurtzberg at Duke University Medical Center. Both transplants are successful and patients suffer remission from their disease for significant periods of time.
- ***January 22:*** Newly elected president Bill Clinton issues a memorandum revoking the existing ban on fetal and embryonic research.
- ***June 10:*** Congress passes the NIH Revitalization Act of 1993 (later Public Law 103-43, Section 121(c)), effectively nullifying existing prohibitions against many forms of research on human embryos and fetuses. One provision of the act requires the National Institutes of Health (NIH) to establish a Human Embryo Research Panel (HERP) to assess the moral and ethical issues raised by in vitro fertilization and related research.

1994

- ***September 27:*** The Human Embryo Research Panel (HERP), charged with developing standards for determining which fetal and embryonic research projects can ethically be supported with federal funds and which cannot, issues a report, listing areas of research that it regards as acceptable and unacceptable. Among the approved activities is the creation of human embryos specifically for the purpose of experimentation.
- ***November:*** Sri Lankan gynecologist Ariff Bongso obtains stem cells from a five-day-old embryo and demonstrates that they are able to differentiate in vitro to produce other types of cells. Bongso is able to keep the embryonic stem cells alive and reproducing for two generations on the surface of fallopian tube tissue, but is unable to extend their lives beyond that point. Questions remain as to whether the cells obtained by Bongso were pure embryonic stem cells or whether they were cells that had already begun to differentiate.

- ***December 2:*** President Bill Clinton issues an executive order directing the National Institutes of Health (NIH) not to provide funding for any research that would involve "the creation of human embryos for research purposes," although the use of "spare embryos," donated by individuals who give informed consent is permitted. Clinton's order directs NIH to develop guidelines for the use of "spare embryos" in research.

1995

- Representative Jay Dickey (R-Fla.) offers a rider to the appropriations bill for the Department of Health and Human Services (HHS). The Dickey amendment prohibits the use of federal funds for the creation of human embryos for research or for research that makes use of human embryos that have been destroyed.
- ***October 3:*** President Bill Clinton issues Presidential Executive Order 12975 creating the National Bioethics Advisory Commission (NBAC) to provide guidance to federal agencies on the ethical conduct of current and future human biological and behavioral research. The commission has since been replaced by the President's Council on Bioethics, created in connection with President George W. Bush's address on human stem cell research of August 9, 2001.

1996

- ***January 26:*** The Dickey amendment is passed by Congress. A similar rider is adopted in each subsequent year through 2001, after which President George W. Bush's statement of policy on human embryo stem cell research makes the prohibition, to a large extent, moot.
- ***July 5:*** A sheep named Dolly is born at the Roslin Institute in Scotland as the result of a cloning experiment conducted by a research team led by Ian Wilmut. Dolly is euthanized in 2003 as the result of a lung disorder believed to have been caused by her atypical method of conception.

1997

- ***February 24:*** Researchers at the Roslyn Institute in Edinburgh, Scotland, led by Ian Wilmut, announce that they have used the process of somatic cell nuclear transfer to produce the first cloned mammal, a sheep named Dolly. The announcement comes many months after the sheep's birth.
- ***February 24:*** President Bill Clinton asks the National Bioethics Advisory Commission (NBAC) to review the ethical and legal issues associated with the use of cloning technology. The commission issues its report on June 9, 1997, recommending that the existing moratorium on the use of

federal funding for somatic cell nuclear transfer research be continued and that researchers being funded by private funds be asked to comply voluntarily with the intent of the federal moratorium.

- ***July:*** John Gearhart, from Johns Hopkins University, announces that he and his colleague Michael Shamblott have isolated germ cells from a human fetus. He also says that their research team is in the process of proving that the cells are pluripotent, an experiment that results in their historic paper of November 10, 1998, published in *Proceedings of the National Academy of Sciences.*

1998

- ***January 7:*** Independent scientist Dr. Richard Seed announces his intention to clone a human being.
- ***March 6:*** A group of Italian researchers led by Giuliana Ferrari discovers that hematopoietic stem cells are able to migrate to muscular tissue, where they differentiate and develop into normal muscle cells.
- ***November 6:*** A research team at the University of Wisconsin–Madison led by James A. Thomson announces the first isolation of human embryonic stem cells from a group of 36 fertilized eggs that had been grown to the blastocyst stage before being sacrificed to permit removal of the inner cell masses. Five of the stem cells are cultured into cell lines that continue to grow for periods of four to five months.
- ***November 10:*** In research closely related to that of Thomson, a research team from the Johns Hopkins University led by John Gearhart announces the isolation of human embryonic germ cells. The Johns Hopkins research differed from the Wisconsin research in two respects. First, the stem cells isolated were germ cells, which, after differentiation, become egg and sperm cells, not somatic cells, as in the case of the Thomson experiment. Second, the cells were obtained from gonadal ridges and mesenteries (membranous tissue that surrounds, supports, and carries blood to the intestines) of fetuses five to nine weeks old (postfertilization).

1999

- ***January 15:*** Department of Health and Human Services General Counsel Harriet Rabb rules that the Dickey amendment restricting the funding of stem cell research on embryos does not prohibit the federal funding of research on stem cells obtained from embryos produced through private funding.
- ***January 22:*** A research team of Canadian and Italian scientists headed by graduate student Christopher R. R. Bjornson announces that neural stem cells from mice are capable not only of generating new neural cells,

but also of differentiating into certain types of blood cells, including myeloid, B, and T leukocytes.

2000

- ***June 18:*** A committee headed by Sir Liam Donaldson, Chief Medical Officer of Great Britain, issues a report on recent developments in in vitro fertilization and other forms of assisted human reproduction. The committee essentially confirms the general policies established by the 1990 Human Fertilisation and Embryology Act.
- ***December 1:*** The journal *Science* carries an article written by Éva Mezey, researcher at the U.S. National Institutes of Health (NIH) reporting the conversion of hematopoietic cells into neurons in rat brains. The study is so revolutionary and controversial that the magazine holds the report for over a year before allowing publication. The report accompanies a second paper by a group of researchers at Stanford University led by graduate student Timothy Brazelton reporting similar results with hematopoietic bone marrow cells that migrate to the brain and differentiate to produce new neuronal cells.

2001

- ***May 4:*** A research team headed by Diane Krause, at the Yale University School of Medicine, reports that a single hematopoietic bone marrow cell is able to differentiate into many different types of specialized cells, such as those found in the digestive system, the liver, the lungs, and the skin.
- ***July 31:*** The U.S. House of Representatives passes house bill H.R. 2505, the Human Cloning Prohibition Act of 2001, by a vote of 265-162. The bill bans the production of human embryos for any reason whatsoever, including both reproductive purposes and research on stem cells for therapeutic purposes. A companion bill, S. 1899, was introduced into the Senate, but never acted upon. Both bills became moot when President George W. Bush announced in August federal policy on stem cell research.
- ***August 9:*** In a nationally televised address, President George W. Bush outlines his administration's policy on stem cell research. He will permit research on certain preexisting human embryonic stem cell lines, on adult stem cells, and on cord blood stem cells, but will not permit the creation of new embryos for the purpose of either therapeutic research or human cloning experiments.
- ***November 25:*** Advanced Cell Technology (ACT), a biotechnology research company located in Worcester, Massachusetts, announces two breakthroughs in studies of embryonic stem cells. In the first line of re-

search, ACT scientists are successful in coaxing oocytes (egg cells) to begin reproduction without the addition of a male gamete (sperm cell), a process known as parthenogenesis. The activated oocyte is allowed to grow to the preimplantation stage. In the second line of research, DNA from both an oocyte and a sperm cell are removed, and the DNA from a somatic cell is transplanted into the oocyte nucleus. The oocyte is then stimulated to begin dividing and allowed to develop to the 16-cell stage before being sacrificed.

- ***November 28:*** President George W. Bush establishes the President's Council on Bioethics by Executive Order 13237.
- ***December 4:*** The British Parliament passes the Human Reproductive Cloning Act, legislation that makes it a criminal offense to implant a cloned embryo into a woman. The act was made necessary when a British court ruled in 2000 that an earlier law, the Human Fertilisation and Embryology Act of 1990, did not specifically prohibit the cloning of a human being.
- ***December 22:*** Scientists at Texas A&M College of Veterinary Medicine announce the birth of the first cloned cat, a kitten they name CC (for "carbon copy").

2002

- ***February:*** Two researchers from Boston's Whitehead Institute for Biomedical Research, Konrad Hochedlinger and Rudolf Jaenisch, report on the cloning of mice using stem cells taken from the stomach of adult mice. The report is the first occasion on which the successful use of adult stem cells for cloning of an animal has been announced.
- ***February:*** Researchers at Geron Corporation led by Chunhui Xu announced that they have developed a method for maintaining stem cell lines without the use of feeder layers of any kind. The discovery is of importance because of long-term concerns that contaminants in traditional kinds of feeder layers, either murine or human, might become incorporated into the stem cells on which they were maintained, adding a problem for the use of such cells in regenerative medicine.
- ***April:*** A second research team at the Whitehead Institute reports on an experiment in which mice with an immune disease are cured by the transplantation of genetically engineered embryonic stem cells. The experiment was the first confirmed example of the successful therapeutic use of embryonic stem cell therapy.
- ***July 4:*** A research team led by Belgian-born hematologist Catherine Verfaillie, now head of the University of Minnesota's Stem Cell Institute, announces the discovery of a specialized kind of mesenchymal stem cell

with the ability to differentiate into cells characteristics of all three blastocytal layers, epidermal, mesodermal, and endodermal. The researchers note that, if their findings are confirmed, these mesenchymal cells "May be an ideal cell source for therapy of inherited or degenerative diseases."

- ***September 22:*** Governor Gray Davis of California signs a bill passed by the state house and senate amending portions of the California Health and Safety Code authorizing the use of human embryonic stem cells, human embryonic germ cells, and human adult stem cells obtained from any source, as well as the process of somatic cell nuclear transplantation for research and therapeutic purposes, but not for the purpose of human reproductive cloning. The bill also requires health care providers to inform individuals whom they treat and advise of the possibility of donating surplus embryos produced in their treatment for stem cell research.

2003

- ***May 15:*** Eleven Republican members of the U.S. House of Representative write to President George W. Bush, expressing their concerns over the progress (or lack of progress) of stem cell research at the National Institutes of Health (NIH) and asking that the president revisit his policy on stem cell research. The president declines to do so.

2004

- ***January:*** The governments of both China and South Korea adopt legislation prohibiting reproductive cloning, but specifically allow human embryo cloning for the purpose of research.
- ***January 4:*** New Jersey Governor James E. McGreevey signs into law senate bill 1909, the Stem Cell Research Bill, legalizing stem cell research in the state and providing mechanisms for its funding with state tax monies. New Jersey becomes the second state, after California, to legalize stem cell research.
- ***February 12:*** South Korea researchers at Hanyang University under the direction of Hwang Woo Suk of Seoul National University announce that they have produced 30 human embryos by means of somatic cell nuclear transfer. The embryos were produced from 242 eggs taken from 16 volunteers and were allowed to reach the blastocyst stage before being sacrificed. (See May 2006.)
- ***May 19:*** The British government announces the opening of the world's first embryonic stem cell bank. The bank expects to receive, store, and supply tens of thousands of stem cell lines to be used in research for the treatment of chronic diseases, such as diabetes, Alzheimer's disease, and Parkinson's disease.

- ***October:*** The Virginia Commonwealth University Life Sciences Survey releases the results of its most recent nationwide survey of public opinion on stem cell research. The survey finds that support for embryonic stem cell research has increased to 53 percent of those surveyed, compared to an approval rating of 47 percent only a year earlier. At the same time, opposition has dropped from 44 percent to 36 percent during the preceding year.
- ***November 2:*** Voters in the state of California pass Proposition 71 by a margin of 59 percent to 41 percent. The proposition creates a new California Institute for Regenerative Medicine, empowered to regulate stem cell research and to provide funding for such research and necessary research facilities. The proposition also establishes a constitutional right for the conduct of stem cell research in the state and specifically prohibits state funding for any form of human reproductive cloning research. The proposition also authorizes the expenditure of $3 billion over a 10-year period for stem cell research in the state.

2005

- ***February 16:*** Representatives Mike Castle (R-Del.) and Diana DeGette (D-Colo.) introduce a bill into the U.S. House of Representatives (H.R. 810) that authorizes the use of federal funds for stem cell research using materials taken from surplus embryos that have been donated for that purpose, provided that certain specific conditions have been met for the donation. Human reproductive cloning is specifically prohibited by the bill. The bill is passed by the full house on May 24, 2005.
- ***March 4:*** The Brazilian Congress passes and President Luiz Inacio Lula da Silva later signs a Bio Safety Law that authorizes the conduct of embryonic stem cell research in the country.
- ***March 17:*** Dave Weldon (R-Fla.) introduces a bill into the U.S. House of Representatives to prohibit all forms of human cloning, for whatever purpose, including both the reproduction of new humans and the production of embryos for stem cell research. It also prohibits the shipment, receipt, exportation, or importation of any human embryo or any product from a human embryo that has been produced by cloning. The bill is only one, but probably the best known, of similar bills seeking to reinforce the federal government's ban on stem cell research using embryos or embryo products. The bill is never acted upon by the House, as is the case with almost all other such bills.
- ***May 19:*** The Massachusetts state legislature overrides the veto of Governor Mitt Romney to a bill that would permit the creation of embryos for scientific research. Massachusetts becomes the third state, after

California and New Jersey, to pass such a law. The bill passes 35-2 in the Massachusetts senate and 112-42 in the house.

- ***May 20:*** South Korean scientists announce that they have, for the first time, produced 11 new stem cell lines tailored to match a specific individual (this later turns out not to be true). The scientists use the process of somatic cell nuclear transfer (SCNT) to insert the nuclei from skin cells taken from a person to be treated into an enucleated egg from a donor. (See May 2006.)
- ***May 23:*** Representative Christopher H. Smith (R-N.J.) introduces a bill into the U.S. House of Representatives requiring the Secretary of Health and Human Services (HHS) to enter into agreements with organizations that collect and store cord blood stem cells, with the goal of collecting and maintaining 150,000 units of human cord blood. The blood is to be made available for transplantation through an existing program of HHS, the C. W. Bill Young Cell Transplantation Program. The bill is passed by a vote of 431-1 in the House on the next day and forwarded to the Senate.
- ***June 15:*** Connecticut Governor Jodi Rell signs legislation passed by both houses of the Connecticut legislature authorizing the expenditure of $100 million in state funds to finance stem cell research in the state.
- ***July 12:*** Illinois governor Rod Blagojevich sets aside $10 million in state funds to support stem cell research. The governor acted after a bill designed to achieve the same objective failed in the state legislature in fall 2004, largely as the result of vigorous opposition by religious groups.
- ***July 15:*** A CBS News poll of 632 adults nationwide finds that 58 percent of respondents approve of "medical research using embryonic stem cells," while 30 percent disapprove.
- ***October 13:*** A study conducted by researchers from the Genetics and Public Policy Center at Johns Hopkins University finds that 67 percent of those Americans interviewed either approve or strongly approve of stem cell research. Forty percent favor the expansion of government support for embryonic stem cell research. By contrast, 16 percent of respondents would ban all research using embryonic stem cells, and 22 percent support the Bush administration's policy on stem cell research.
- ***October 15:*** University of Minnesota Stem Cell Institute researcher Dan Kaufman announces that his research team has found a way to coax human embryonic stem cells to differentiate into natural killer (NK) cells that seek out and destroy cancer cells.
- ***November:*** A group of business leaders, patient advocates, and researchers and Missouri file an initiative petition to permit cloning of human eggs for research purposes. The initiative is opposed by religious leaders who file suit to oppose the plan and, in some cases, preach against the proposal in local churches throughout the state.

- ***November 25:*** South Korean researcher Hwang Woo Suk apologizes for ethical errors in the historic research on stem cells conducted in his laboratory. Some of the eggs used in the South Korean's research were extracted from two junior scientists of the team. He announces that he will resign his post at Seoul National University.
- ***December:*** Reaction to Hwang Woo Suk's announcement of November 25 in South Korea is overwhelming, with government officials, scientists, and ordinary citizens announcing their support for the scientist, who has become a great hero for his work on stem cells. Supporters plead for his return to the research laboratory, and more than 700 women pledge to donate eggs for his future research.
- ***December 14:*** Gerald Schatten, coauthor of the May 20, 2005, report on 11 new stem lines produced in South Korea, asks that the journal *Science*, publishers of the report, remove his name from the report because of ethical issues involved in conduct of the research about which he was not aware.

2006

- ***February 3:*** New York City Mayor Michael Bloomberg donates $100 million of his own money to Johns Hopkins University, part of which is allocated to the funding of stem cell research.
- ***February 10:*** The United Kingdom's Human Fertilistion and Embryology Authority announces that it will change current rules to allow any woman who chooses to donate eggs for use in stem cell research. Currently, only women undergoing in vitro fertilization treatment are allowed to make such donations.
- ***February 16:*** The National Academy of Sciences establishes a committee to provide informal oversight for stem cell research. Because of the Bush administration's objections about stem cell research, the committee will be funded by private organization rather than with federal tax dollars.
- ***February 27:*** Three organizations opposed to stem cell research—People's Advocate, National Tax Limitation Foundation, and California Family Bioethics Council—file suit in Alameda (California) County Superior Court to prevent implementation of Proposition 71, adopted in the state's November 2004 election. The suit is later dismissed.
- ***March 10:*** Oregon Health Sciences University announces that it will use stem cells harvested from human fetuses to treat six children with Batten disease, a neurological disorder that is always fatal. It is the first research of its kind in the world.
- ***March 13:*** The European Union fails to agree on a common stem cell research policy. Policies vary throughout the Union, ranging from total bans on such research in some nations to legal and/or financial support

for stem cell research in others. Officials had hoped to reach a compromise that could be adopted throughout the Union but were unsuccessful in achieving that agreement.

- ***May 12:*** Prosecutors in South Korea indict Hwang Woo Suk for faking most of his stem cell research.
- ***July 18–19:*** By a vote of 63-37, the U.S. Senate passes S. 471, companion bill to H.R. 810, which was passed by the House of Representatives in 2005. A day after Senate passage, President George W. Bush vetoes the bill, exercising his veto power for the first time in nearly six years as president. Bush says the bill crosses "a moral boundary," a move that he would not accept. The House of Representatives votes to override the veto, but there are not sufficient enough votes to do so, therefore the veto remains.

CHAPTER 4

BIOGRAPHICAL LISTING

This chapter contains brief biographical sketches of individuals who have played major roles in stem cell research and issues that have developed with respect to that practice.

Joseph Altman, currently professor emeritus at Purdue University, in Lafayette, Indiana. Altman carried out seminal studies on the production of new brain cells in the 1960s. Prior to Altman's work, nearly all scientists believed that a person was born with all of the brain cells it would ever have. Altman demonstrated that he observed new neurons (brain cells) being created in the hippocampus portion of the brain and then migrating to other parts of the brain. Altman's studies were largely ignored for at least a decade by many scientists, and for even longer by the majority of the scientific community.

Elizabeth Blackburn, Morris Herzstein professor of biology and physiology in the department of biochemistry and biophysics at the University of California–San Francisco. She is a widely respected authority in the field of stem cell research. Her field of expertise is telomeres, regions located at the ends of chromosomes that control the process of replication. She discovered an enzyme known as telomerase that controls this process. Blackburn served on the President's Bioethics Advisory Commission from 2002 to 2004 before being asked to resign by President George W. Bush. The official reason offered for this action was that Blackburn had been unable to attend meetings of the commission. Blackburn and others, however, said that their views on embryonic stem cell research were regarded as too liberal for the administration. She has been critical of the operation and activities of the commission.

Rod Blagojevich, governor of Illinois. Blagojevich signed an executive order on July 12, 2005, committing $10 million in state funds over the coming year for research on stem cells. Blagojevish pointed out that the decision by the federal government not to fund embryonic stem cell research made it imperative for individual states to take over part of that funding responsibility. Prior to his election as governor, Blagojevich served as member of the state assembly from 1992 to 1996 and in the U.S. House of Representatives from 1996 to 2002.

Ariff Bongso, research professor in the department of obstetrics and gynaecology at the faculty of medicine of the National University of Singapore. He was the first person to isolate human embryonic stem cells from a five-day-old human embryo in 1994. Bongso was trained as a veterinarian in his native Sri Lanka but later became interested in the young science of in vitro fertilization (IVF), where he made a number of research breakthroughs while working in Singapore, to which he had moved in the early 1980s. He was a member of the IVF team that produced the first successful "test-tube baby" born in Asia in 1983. In addition to his isolation of the first human embryonic stem cells, Bongso has developed methods for maintaining human embryonic stem cells in a regenerating state outside the human body for essentially limitless periods of time, allowing the establishment of new stem cell lines.

Robert Briggs, former head of the embryology department at the Institute for Cancer Research and professor of zoology at Indiana University. With colleague Thomas J. King, he carried out some of the earliest research on somatic cell nuclear transfer, in which the nucleus of a cell from one organism is transplanted into an egg from which the nucleus has been removed of a second organism.

Ralph Brinster, currently Richard King Mellon Professor of Reproductive Physiology at the University of Pennsylvania's college of veterinary medicine. He was responsible for some of the earliest research on the fate of nonembryonic stem cells transplanted into mouse blastocysts. He transplanted cells taken from teratocarcinomas and the bone marrow of mice into the blastocysts of host mice and found that those blastocysts grew into normal adult mice with the characteristics of both the host mouse blastocysts and the mice from which the transplanted cells had been taken.

Sam Brownback, Republican senator from Kansas. He has been one of the strongest opponents of stem cell research that results in the destruction of embryos or other early forms of life in the U.S. Congress. He has introduced bills to prevent cloning for either reproductive or therapeutic purposes in every session of Congress in recent years. Two such bills were S. 245 and S. 658, the Human Cloning Prohibition acts of 2003 and 2005, respectively.

George W. Bush, 43rd president of the United States. Bush announced his administration's policy on stem cell research in an address to the nation on August 9, 2001. That policy allowed the use of certain existing stem cell lines for research on human embryonic stem cells, but prohibited federal funding of research in which new embryos would be created solely for the purpose of experimentation. Since his 2001 address, the president has maintained his opposition to any type of stem cell research in which embryos are created and then destroyed for any purpose whatsoever.

Lisa Sowle Cahill, J. Donald Monan professor of theology at Boston College. She has written extensively on ethical issues related to stem cell research, cloning, abortion, and other bioethical issues. Some of her most important work has involved the commercialization of new biotechnological procedures, such as those concerned with the production and use of human embryos for stem cell research.

Daniel Callahan, cofounder and director of the International Program at the Hastings Center, a bioethics research institution in upstate New York. He is a highly respected observer of and commentator on the development of stem cell research. He has been opposed to embryonic stem cell research because, as he has explained, he has "always felt a nagging uneasiness at trying to rationalize killing something for which I claim to have profound respect [a human embryo]." He also spoke out in opposition to California's Proposition 71, establishing the California Institute of Regenerative Medicine to be funded at a level of $3 billion over a 10-year period. "Whether anything comes of this research," he said, "it is sure to line the pockets of many scientists and biotechnology companies in the process."

Mike Castle, Republican representative from the state of Delaware. Castle was cosponsor of House bill H.R. 810 in the 109th Congress. The bill authorized the Secretary of Health and Human Services to "conduct and support research that utilizes human embryonic stem cells." The bill passed the House, was never acted on by the Senate. Castle was previously deputy attorney general, state legislator, lieutenant governor, and, for two terms, governor of Delaware. He was elected to the House of Representatives in 1993.

William Jefferson Clinton, 42nd president of the United States. Clinton was confronted with growing interest in and conflict about the use of embryonic stem cells in research throughout his presidency. He inherited a policy that banned the use of all federal funds for such research from his predecessor, President George H. W. Bush, but gradually liberalized that policy in order to permit the use of at least certain types of embryonic stem cells under certain conditions. By the time he left office in 2001, Clinton had set in motion a plan to permit the funding of such research with tax dollars, a policy that was reversed shortly after his successor, George W. Bush, took office in January 2001.

Diana DeGette, Democratic representative from the state of Colorado. She was cosponsor of House bill H.R. 810 in the 109th Congress, a bill authorizing the Secretary of Health and Human Services to conduct and support research using human embryonic stem cells. She served two terms in the Colorado legislature before being elected to the U.S. House of Representatives in 1997.

Jay Dickey, formerly a congressman from the 4th congressional district of Arkansas. He was author of an amendment to the annual appropriations bill for the Department of Health and Human Services in 1996, an amendment that prohibits the allocation or expenditure of any federal funds for research that involves the destruction of a human embryo or embryos. That amendment has generally become known as the Dickey amendment and has been reintroduced and adopted every year since it was first passed in 1996. Dickey was first elected to Congress in 1992, where he served for four terms until his defeat in 2000.

Richard M. Doerflinger, deputy director of the Secretariat for Pro-Life Activities of the United States Conference of Catholic Bishops. He is a prominent spokesperson in opposition to the use of embryos and materials obtained from embryos in research of any kind whatsoever. He is also adjunct fellow in bioethics and public policy at the National Catholic Bioethics Center in Boston. Doerflinger has written extensively for the Hastings Center Report, the Kennedy Institute of Ethics Journal, the Encyclopedia of Catholic Doctrine, the Dusquene Law Review, and a number of Catholic magazines and journals. He has also testified before the U.S. Congress, the National Bioethics Advisory Commission, and the National Institutes of Health on issues related to the use of human embryos in research.

Hans Driesch, German biologist and philosopher, carried out a series of classic experiments in the early 1890s in which he divided two- and four-cell sea urchin embryos and found that each of the individual cells produced was able to develop into a complete, normal adult sea urchin. In order to explain his results, Driesch later developed a theory of "entelechy" that attributed an organism's growth and development to some sort of supernatural "unifying non-material mind-like something." The effect observed by Driesch was later found to result from the proliferation and differentiation of embryonic stem cells.

Robert Edwards, professor emeritus at the University of Cambridge, England. With colleague Barry Bavister, he performed the first successful ex utero fertilization of a human egg in 1968. The experiment was one of the seminal steps in developing the procedure of in vitro fertilization, which has since been responsible for the birth of tens of thousands of children to otherwise infertile couples.

Sir Martin Evans, professor of mammalian genetics and director of the school of biosciences at Cardiff University, in Wales. Along with colleague

Matthew Kaufman, he was the first person to successfully isolate embryonic stem cells from a mouse. He and Kaufman then developed a process by which those stem cells could be kept alive in a proliferative, nondifferentiating state for many generations, producing the first embryonic stem cell lines available for research. Because of this research, Evans has sometimes been called "the chief architect of stem cell research." In addition to his work on stem cells, Evans has long been interested in the development of modern gene therapy, in which altered genes are introduced into organisms suffering from some sort of genetic disorder. During the 1990s, his research team performed the first successful experiments in curing a mouse with cystic fibrosis by means of gene therapy.

John D. Gearhart, C. Michael Armstrong professor of medicine and professor of gynecology and obstetrics, physiology, and comparative medicine at the Johns Hopkins School of Medicine and professor of biochemistry and molecular biology at the Johns Hopkins School of Public Health and Hygiene. Gearhart published an historic paper in stem cell research in 1998 when he reported on the derivation of human embryonic stem cells from primordial germ cells. Gearhart also serves as director of the Division of Development Genetics, director of Research for Gynecology and Obstetrics, and director of Preimplantation Genetic Diagnosis at the Johns Hopkins School of Medicine. In addition to his work on stem cells, Gearhart has published papers in the fields of genetics, development, and genetic counseling.

Howard Green, currently George Higginson Professor of Cell Biology at Harvard University. He invented a method for growing cells in vitro on irradiated mouse fibroblast cells. Green's murine fibroblast mat was later to become the standard feeder layer on which stem cells are maintained in a proliferative, nondifferentiating state over periods of many months or years. In later research, Green found ways to embed epidermal stem cells in fibroblast mats for use as synthetic skin for patients who had been severely burned, earning him the accolade from some as Father of Skin Culture.

Jim Greenwood, president of the Biotechnology Industry Organization. He represented the 8th congressional district of Pennsylvania from 1993 to 2004. During his tenure in office, Greenwood was a strong supporter of embryonic stem cell research and introduced a number of bills authorizing the funding of such research with federal tax monies. As an example, he cosponsored (with representative Peter Deutsch) H.R. 2608 in the 107th Congress, a bill that would have permitted the production and use of human embryos for research and therapeutic purposes, but not for human reproduction. Greenwood's bills consistently lost out to more restrictive legislation banning all forms of embryonic research.

John B. Gurdon, research scientist and group leader in the Institute of Cancer and Developmental Biology at Cambridge University. He was the first person to successfully clone an animal. During the 1960s, Gurdon carried out a number of experiments that destroyed the nucleus of the eggs of frogs, and transplanted into those eggs the nuclei from tadpoles. Some small number of those eggs eventually developed into new tadpoles, exact copies (clones) of the organisms from which the nuclei had originally been taken.

Gottlieb Haberlandt, Austrian botanist. A pioneer in modern studies of plant tissue culture, he hypothesized in 1902 that it should be possible to grow a complete mature plant beginning with no more than a single cell taken from that plant. He said that this experiment should be possible because each cell in a plant possesses a "totipotency" that allows it to develop into any one of the types of cells of which the mature plant consists and into which it can grow.

John Hearn, Australian reproductive and developmental biologist. He was formerly director of the Wisconsin Regional Primate Research Center, located at the University of Wisconsin–Madison, where he was responsible for recruiting James Thomson, the first person to isolate human embryonic stem cells in 1998, and to provide him and other researchers with unusually fine laboratory conditions in which to conduct their studies.

Konrad Hochedlinger, a postdoctoral researcher at the Whitehead Institute for Biomedical Research in Boston. He demonstrated in 2002 that a mouse can be cloned from mature, highly differentiated cells taken from an adult animal. Later in the same year, Hochedlinger and his advisor-collaborator Rudolf Jaenisch used embryonic stem cells to cure a mouse of an immune disorder, the first time in which a therapeutic application of stem cell research had been conclusively illustrated.

Robert Hooke, English physicist. He was the first person to explicitly recognize the existence of tiny pocketlike units within living organisms. He gave these units the name cell after the Latin word *cella* meaning "small room."

Rudolf Jaenisch, one of the founding members of the Whitehead Institute for Biomedical Research, in Boston. He was joint author with one of his postdoctoral students, Konrad Hochedlinger, of an important paper in 2002 describing the cloning of mice using adult stem cells, the first time such a procedure had been conclusively shown to work. Jaenisch and Hochedlinger also authored a 2002 paper that described the use of stem cells to cure mice of an immune disorder, the first occasion on which the therapeutic value of embryonic stem cell transplantation had been conclusively demonstrated.

Leon R. Kass, chair of the President's Council on Bioethics, on leave from his positions as Hertog Fellow in Social Thought at the American Enterprise Institute and Addie Clark Harding Professor in Social Thought at the University of Chicago. He earned his Ph.D. in biochemistry at Harvard University in 1967 and was a researcher in that field briefly before turning his attention to ethical and philosophical issues raised by advances in medical research and the biosciences. Since 1970, he has taught and written extensively on a variety of issues in the field of bioethics.

Thomas J. King, professor of embryology at Georgetown University, division director at the National Cancer Institute, director of the Kennedy Institute of Ethics, and deputy director of the Lombardi Cancer Research Center at Georgetown. King collaborated with Robert Briggs in the early 1950s to develop the procedure now known as somatic cell nuclear transfer (SCNT), in which the nuclei are removed from cells of one organism and transplanted into cells that have had their nuclei removed of a second organism. When done successfully, the host cell grows and develops normally, as Briggs and King discovered in their classic studies of the leopard frog (*Rana pipiens*).

Karen Lebacqz, professor emerita of theological ethics at the Pacific School of Religion, Berkeley, California. Lebacqz is coeditor, with Suzanne Holland and Laurie Zoloth, of *The Human Embryonic Stem Cell Debate*, a book of readings on the scientific, legal, and ethical issues related to stem cell research. She has also written, spoken, and taught extensively on other areas of bioethics, including bioethics of the Human Genome Project and ethical theory.

Gail Martin, professor of anatomy at the University of California at San Francisco. She is codiscoverer with Martin Evans and Matthew Kaufman of murine (mouse) embryonic stem cells in 1981. She is often credited with having invented the terminology now in common use for these cells. Her current research focuses on the roles played by specific molecules in the early differentiation of murine embryonic stem cells.

Alexander A. Maximow, a Russian military officer who first proposed the notion that all kinds of blood cells—white blood cells, red blood cells, and platelets—are all produced within bone marrow from a single precursor cell, which he called a Stammzelle. Some scholars believe that the modern term *stem cell* can be traced to Maximow's use of the similar term *Stammzelle.* Maximow's ideas were largely rejected or ignored throughout his lifetime and received experimental justification only with the work of Ernest McCulloch and James Edgar Till in the 1960s.

Richard McCormick, a moral theologian who taught at Georgetown University. He wrote extensively about the moral status of the human embryo, with special attention to its potential use in human embryonic stem cell research. After considerable research and deliberation, McCormick came to the conclusion, contrary to church doctrine, that the very early embryo was not truly a human and that, therefore, an argument could be made for its use in experimentation, provided the end results of such experimentation justified that use.

Ernest McCulloch, Canadian medical researcher. He carried out pioneering work on hematopoietic stem cells in the 1960s with James Edgar Till and Andrew Becker that provided experimental evidence for the existence of such cells, produced methods for the culturing of stem cells, and found methods for counting the number of hematopoietic cells in bone marrow and other body organs.

Douglas Melton, Thomas Dudley Cabot professor of the natural sciences at Harvard University and codirector of the Harvard Stem Cell Institute. His research focuses on the development of the pancreas, in general, and, more specifically, on the role played by stem cells in that process. One product of his research has been the development of 17 new stem cell lines, produced with private funding. Melton has testified before the U.S. Congress in support of federal funding for stem cell research and is active in a number of organizations promoting the conduct and funding of stem cell research.

Éva Mezey, Hungarian-born medical researcher, current head of the Adult Stem Cell Research Section at the National Institute of Dental and Craniofacial Research of the U.S. National Institutes of Health. In the late 1990s, Mezey discovered that some of the hematopoietic cells transplanted into mice migrated to the brain and transdifferentiated into neurons. Her discoveries were, at first, largely disbelieved, although many similar cases of transdifferentiation have since been discovered.

Beatrice Mintz, senior member of the basic science division of Fox Chase Cancer Center in Philadelphia. She carried out an elegant series of experiments in the early 1970s in which stem cells taken from a teratoma were transplanted into a normal blastocyst. The cells were incorporated into the blastocyst and became part of the normal embryo into which it grew. Mintz is perhaps best known for her work resulting in the production of the first transgenic mammals, organisms produced when the genes from one mouse have been transplanted into the body of a second mouse, resulting in an animal with two different genetic maps.

Thomas Okarma, current president and chief executive officer of Geron Corporation. After serving as a faculty member at the Stanford University school of medicine, Okarma joined the corporate world, where he was a founder, vice president for research, and president and chief executive officer of Applied Immune Sciences, Inc., and senior vice president at Rhône-Poulenc Rorer before joining Geron in 1997. At Geron, Okarma was vice president of cell therapies and vice president of research and development before being appointed to his present posts at the company.

Geron funded Dr. James Thomson's research on stem cells and has been awarded patents for nine stem cell lines developed at the University of Wisconsin–Madison.

Deborah Ortiz, state senator for California's 6th state senate district. She has been an outspoken advocate for all forms of stem cell research and was the author of senate bill 253 in the 2002 legislative session, affirming the state's support of human embryonic stem cell research. She was also active in the campaign for Proposition 71 in the November 2004 election, an initiative that created the California Institute for Regenerative Medicine, to be funded with $3 billion of taxpayer monies over a 10-year period.

Gordon Barry Pierce, former professor of pathology at the University of Colorado Health Sciences Center in Denver. He was codiscoverer with medical student Lewis Kleinsmith that teratomas are caused by the proliferation of individual stem cells known as embryonal cancer (EC) cells. Born in Canada, Pierce spent virtually all of his professional career in the United States. He is best known for his studies of the character of teratomas and the role of EC cells in their development.

Harriet Rabb, current vice president and general counsel to the Rockefeller University. She wrote a critical memorandum in 1995 on the legality of federal funding for human embryonic stem cell research while she was general counsel for the U.S. Department of Health and Human Services. Rabb ruled that, while federal money could not be used for the production of human embryos for research, it could be used for research on such embryos produced with private funds.

Nancy Reagan, wife and widow of former president Ronald Reagan. She became an outspoken advocate of all forms of stem cell research during the later stages of her husband's battle with Alzheimer's disease. At one point, she announced that "we have already waited too long" to find out how stem cells might be useful in treating diseases like the one from which her husband was suffering. During the 2005 congressional debate over federal funding of embryonic stem cell research, she encouraged both President George W. Bush and members of congress to act to approve federal funding for such research.

Ron Reagan, Jr., son of former president Ronald Reagan. He is a strong advocate for stem cell research, arguing that it has the potential for curing a broad range of debilitating diseases currently untreatable by other methods. He spoke at the Democratic National Convention in Boston in 2004, arguing that, although some people opposed embryonic stem cell research because they believed that embryos are live human beings, "[i]t does not follow that the theology of a few should be allowed to forestall the health and well-being of the many."

Christopher Reeve, an actor with credits in stage, screen, and television productions. Reeve will perhaps always be best known for his portrayal of Superman in the film of that name and its sequels. In 1995, he was thrown by the horse he was riding in an equestrian competition and paralyzed as a result of an injury to his spine. During the next decade, he became very active in efforts to promote research for the treatment and cure of spinal cord injuries, including stem cell research. He died on October 10, 2004, although his work in the support of medical research is being carried on by the Christopher Reeve Paralysis Foundation.

Matthias Jakob Schleiden, 19th-century German botanist. He was the first to hypothesize that cells are the structural units of which all plants are made. His suggestion, along with a similar theory for the structure of animals by Theodor Schwann, constitute the basis of modern cell theory. At one point, Schwann makes the prescient observation that "any given cell may be separated from the plant, and then grown alone."

Theodor Schwann, 19th-century German physiologist. He theorized that all living organisms are made of cells and that these cells grew out of pre-existing cells. Schwann also discovered the enzyme pepsin, hypothesized that fermentation was a biological process, and identified yeast cells as plantlike organisms.

Patrick Steptoe, British obstetrician and gynecologist. With colleague Robert Edwards, he performed the first successful in vitro fertilization of a human egg that resulted in the birth of a normal child, so-called "test tube baby" Louise Brown, in 1978.

Leroy Stevens, long-time researcher at the Jackson Laboratory, in Bar Harbor, Maine. He has been called "the unsung hero of stem cell research" because of his early discovery of pluripotent cells in mice. Stevens's research on pluripotent cells, a term he invented, began shortly after he joined the Jackson Laboratory in 1952 when he discovered the presence of teratomas, tumors consisting of many different kinds of cells, in the scrotums of a particular line of experimental mice. He spent the rest of his life studying these cells, retiring from the Jackson Laboratory in 1989.

Frederick Campion Steward ("Camp" Steward), former director of the laboratory of cell physiology, growth, and development at Cornell University. He became famous for a series of experiments he and his colleagues conducted in the 1950s during which they were able to regenerate a complete carrot plant beginning with a single carrot cell cultured in coconut milk. The experiment demonstrated the possibility of dedifferentiating a mature cell into a more primitive form that was then able to reproduce and form all of the mature cells needed to form a complete plant.

Tommy Thompson, 19th secretary of Health and Human Services. He was previously a member of the Wisconsin state assembly and governor of Wisconsin for four terms, from 1987 to 2001. He served in the George W. Bush administration from 2001 until his resignation in 2005. Prior to his appointment as secretary of Health and Human Services and during his first six months in office, he was an ardent supporter of stem cell research and encouraged Congress to support funding of SCR. After President Bush's speech on August 9, 2001, outlining his opposition to most forms of stem cell research, however, Thompson changed his views and mounted a vigorous support of the president's program.

James Thomson, professor of anatomy at the University of Wisconsin Medical School–Madison. He was leader of a research team that announced the first successful culturing of human embryonic stem cells in a paper published in the journal *Science* in 1998. Thomson received his doctoral degree in veterinary medicine from the University of Pennsylvania in 1985 and his Ph.D. in molecular biology from Penn in 1988. His current research studies factors that promote the self-renewal of stem cells, the maintenance of pluripotency, and the pathways leading to the differentiation to specific cell types.

James Edgar Till, Toronto-born specialist in radiation biology. With Ernest McCulloch and Andrew Becker, he cooperated to produce groundbreaking research on hematopoietic cells in the 1960s. Along with McCullouch, he was inducted into the Canadian Medical Hall of Fame for his work in this area in 2004.

Harold Varmus, current president and chief executive officer of Memorial Sloan-Kettering Cancer Center in New York City, was director of the National Institutes of Health from 1993 to 1999, a period during which much fundamental research on stem cells was being conducted. During that period and since, he was and has been an outspoken supporter of all types of stem cell research, suggesting on one occasion that "[i]t is not too unrealistic to say that this research has the potential to revolutionize the practice of medicine and improve the quality and length of life." Varmus was awarded a share of the 1989 Nobel Prize in Medicine or Physiology for his studies of the cellular origins of genes that are responsible for certain types of cancer.

Catherine Verfaillie, a Belgian-born authority on hematology and oncology, and director of the University of Minnesota's Stem Cell Institute. Her current research interests include the nature, development, and treatment of Fanconi and sickle cell anemia; the processes by which human hematopoietic stem cells develop, proliferate, and differentiate; and the properties of adult stem cells, in general. Verfaillie has written extensively on the scientific characteristics of stem cells and their potential therapeutic value, has appeared as an expert witness before the President's Council on Bioethics, and is coeditor of *Handbook of Stem Cells* (2005).

Irving Weissman, Karel and Avice Beekhuis professor of cancer biology and professor of pathology and developmental biology at Stanford University's School of Medicine. He was the first person to isolate any type of stem cell. In 1988, he demonstrated the existence of hematopoietic stem cells in mice and, four years later, repeated his success with human hematopoietic stem cells. In 2000, a research team led by Weissman became the first to find and isolate stem cells in the human nervous system. Weissman has founded two companies to promote research and development on stem cells, SyStemix and StemCells, Inc., and, in 2002, was named director of Stanford's newly created Stanford Institute for Cancer/Stem Cell Biology and Medicine. He has also served as chair of the Panel on Scientific and Medical Aspects of Human Cloning of the National Academy of Sciences.

Dave Weldon, Republican representative from the 15th congressional district of Florida. He has been an outspoken opponent of embryonic stem cell research. During the 109th Congress, he cosponsored with Representative Bart Stupak (D-Minn.) the Human Cloning Prohibition Act of 2005 (H.R. 1357) to prohibit the cloning of human embryos for any purposes whatsoever, either for purposes of reproduction or to obtain embryos for scientific research.

Michael D. West, current chairman of the board, president, and chief executive officer of Advanced Cell Technologies in Worcester, Massachusetts. He has been a major force in providing the funding necessary to conduct stem cell research in the United States without the expenditure of public tax dollars. In 1990, he founded the biotechnology firm Geron Corporation in 1990, where he served as director and senior executive officer until 1998. He then cofounded another biotechnology company, Origen Therapeutics, a company focused on the development of transgenic technologies. In 1999, he was a member of a group that took controlling interest in Advanced Cell Technologies, where he has remained ever since. In the early 1990s, West obtained funding from private sources for the research at James Thomson's and John Gearhart's laboratories that led to the first isolation of human embryonic and fetal stem cells.

Roger F. Wicker, U.S. representative from the 1st congressional district of Mississippi since 1995. He was cosponsor of the Dickey Amendment in 1996 that prohibited the Department of Health and Human Services from funding any type of research involving the production, purchase, or commerce in human embryos.

Ian Wilmut, professor and head of the department of gene expression and development at the Roslin Institute near Edinburgh, Scotland. He was leader of the research team that cloned the first mammal, a sheep, named Dolly, in 1996. Dolly was euthanized in 2003 because of lung problems believed related to her atypical method of conception and birth.

Laurie Zoloth, professor of medical ethics and humanities, and of religion, at Northwestern University's Feinberg School of Medicine. She has written and spoken extensively on the ethical issues related to embryonic and adult stem cell research. Her current research interests involve ethical problems that have arisen as a result of advances in medical technology and research in genetics. She is a member of the National Advisory Council of the National Aeronautical and Space Administration (NASA), NASA's Planetary Protection Advisory Committee, and the executive committee of the International Society for Stem Cell Research. She also serves as chair of the Bioethics Advisory Board of the Howard Hughes Medical Institute.

CHAPTER 5

GLOSSARY

The terms used in discussions of stem cell research are drawn from a variety of fields, including science, technology, engineering, law, philosophy, religion, and business. This chapter provides definitions for some of the most common terms and phrases used in the field of stem cell research. As indicated earlier in this book, terminology is an especially difficult problem in the field of stem cell research. Individuals may, knowingly or not, use terms in nonstandard ways or with meanings that are not familiar to or accepted by others who use the same words. Precise definitions of some terms may not even be generally accepted even by experts in the field. Definitions listed below are necessarily brief and may sometimes benefit by expanded explanations. Researchers are encouraged to consult a number of references in attempting to find definitions for terms that are likely to be most commonly used. Some of the best of these references include the following:

- Committee on Guidelines for Human Embryonic Stem Cell Research, National Research Council. *Guidelines for Human Embryonic Stem Cell Research*. Washington, D.C.: National Academies Press, 2005, pp. 98–104.
- Dirckx, John H. "What's New in Stem Cell Research," *e-Perspectives*, April 2005, pp. 26–29. Available online. URL: http://www.hpisum.com/perspectives/issue50/update.pdf.
- International Society for Stem Cell Research. "Glossary of Stem Cell–Related Terms." Available online. URL: http://tnt.tchlab.org/stemcells/glossary.htm. Accessed on November 8, 2005.
- *Monitoring Stem Cell Research*. Washington, D.C.: President's Council on Bioethics, January 2004, pp. 147–156.
- Stem Cell Network. Available online. URL: http://stemcellnetwork.com. Accessed on November 8, 2005. Glossary is at http://www.stemcellnetwork.ca/guide/glossary.php.

Glossary

- *Stem Cells: Scientific Progress and Future Research Directions.* Washington, D.C.: Department of Health and Human Services, June 2001. Available online. URL: http://stemcells.nih.gov/info/scireport. Look at Appendix F: Glossary and Terms.

abortion An event in which a pregnancy is terminated. Some abortions occur naturally, while others are conducted for health, personal, or other reasons.

adult stem cell An undifferentiated cell found in some specific types of tissue (such as muscle or nerve tissue) with the ability to renew itself and develop into the type of tissue cell in which it is found. Adult stem cells are also called somatic stem cells. Evidence suggests that some types of adult stem cells may be able to differentiate into tissue cells different from themselves, nerve cells from blood stem cells, and muscle cells from nerve blood cells, for example.

allogeneic transplantation The process by which cells, tissues, or organs from one individual are transplanted into a second individual from the same species.

astrocyte A large cell found in nerve tissue.

autologous transplantation The process by which cells, tissues, or organs from an individual are transplanted back into the same person.

blastema A mass of undifferentiated cells from which an organ or body part develops.

blastocoel A cavity in the blastula of the developing embryo.

blastocyst An early stage of an embryo prior to its implantation into the uterine wall, usually made up of about 150 cells consisting of an inner cell wall and inner cavity, and an outer layer of cells, the trophoblast.

blastomere A cell formed in the first stages of embryonic development, after a fertilized egg has undergone division but before a blastocyst has formed.

blastula A hollow ball of cells one cell thick that appears in the early development of an embryo.

bone marrow stromal cell A stem cell that occurs in bone marrow that may develop into a bone, cartilage, fat, or fibrous connective tissue cell.

cell culture The process of growing cells in an artificial medium for the purpose of scientific research.

cell-based therapy A medical procedure in which stem cells are transplanted into a body with the expectation that they will develop into some specific type of cell that will repair damaged cells or augment the number of cells in some specific tissue.

cell division The process by which a single cell divides to produce two new cells.

cell line A collection of cells kept alive in an artificial environment that continues to reproduce itself essentially forever until its fate is changed by some external factor, such as through an experiment.

chimera An organism whose cells are derived from at least two different organisms, such as a mouse and a human. The term comes from an animal in Greek mythology with the head of a lion, the body of a goat, and the tail of a serpent.

clone An organism that is genetically identical to some original cell from which it was originally derived.

co-culture A group of two or more different kinds of cells that have been grown together.

conceptus A term used to describe an organism in its earliest stages of life, that is, as a zygote, an embryo, or a fetus. The term is sometimes used in an attempt to keep discussions "value-free," and avoid talking about the organism as an "unborn child," a "baby," a "human being," or some other term with philosophical or religious context.

cord blood Blood found in the umbilical cord and placenta.

cryopreservation The process of preserving an organic material by lowering its temperature to a very low point. In most in vitro fertilization facilities, fertilized eggs are cryopreserved by being suspended in liquid nitrogen at a temperature of 196°C.

dedifferentiation A process by which a mature cell with specialized structures and functions reverts to a simpler, more primitive state, as when a unipotent adult somatic cell reverts to a simpler pluripotent or totipotent stemlike cell.

differentiation The process by which a primitive unspecialized cell develops into a specialized cell, such as a muscle, heart, or nerve cell. *See also* **directed differentiation.**

diploid cell A cell that has two sets of chromosomes, one set from the father and one from the mother.

directed differentiation The process by which a researcher establishes conditions so as to encourage a stem cell to develop into some specific type of cell.

DNA An acronym for deoxyribonucleic acid, a chemical compound found in the nucleus of all cells that carries instructions for making the proteins of which all cells are made.

ectoderm The outermost layer of the three layers of cells present in an embryo. The ectoderm eventually gives rise to the cells that make up the skin and the nervous system.

embryo A very early stage in the development of an organism. In humans, the term is used to describe the structure that exists from the time of fertilization until the end of the eighth week of gestation.

embryoid body A clump of cells that develops when stem cells aggregate with each other during the process of cell culturing.
embryonal carcinoma cell (EC cell) A cell derived from a teratoma.
embryonic germ cell (EG cell) A cell that occurs in the gonadal ridge portion of an embryo or fetus. Its properties are similar to those of an embryonic stem cell.
embryonic stem cell (ES cell) A primitive undifferentiated cell that occurs in an embryo with the potential for developing into any one of many kinds of tissue cells, such as heart, muscle, liver, nerve, or brain cells.
embryonic stem cell line A group of embryonic stem cells that have been cultured under in vitro conditions and that have been maintained without differentiation for long periods of time, ranging from a few months to many years.
endoderm The innermost layer of the three layers of cells present in an embryo. The endoderm eventually gives rise to the cells that make up the digestive and respiratory systems.
ex utero fertilization Fertilization that takes place outside of the body. Similar to and, in most cases, identical with in vitro fertilization.
feeder layer A group of cells used in a co-culture to maintain pluripotent stem cells. *See also* **co-culture** and **pluripotency.**
fertilization The process by which male and female cells are joined to each other.
fetus A term used to describe an unborn young organism. In humans, the term is used to describe the unborn child from about two months after conception to birth.
gene A unit of heredity that consists of a specific segment of DNA that directs the formation of a protein.
germ cell An egg or sperm cell. Germ cells originate in the inner cell mass of the embryo before migrating outward and beginning to differentiate.
gonad An organ that produces germ cells. For example, the testis produces sperm cells and the ovary produces oocytes, or egg cells.
haploid cell A cell containing only a single set of chromosomes, half the number normally found in a somatic cell. Haploid cells are usually germ cells.
hematopoietic cell transplantation (HCT) Transplantation of cells with blood-forming potential. Such cells are most commonly removed from human bone marrow, but they may also be obtained from umbilical cord blood, the fetal liver, and a few other sources.
hematopoietic stem cell A stem cell found in bone marrow from which all kinds of red and white blood cells eventually develop.
human embryonic stem cell A pluripotent stem cell found in the inner cell mass of the blastocyst of the human embryo. *See also* **pluripotency.**

implantation The process by which the blastocyst is embedded into the endometrium, the lining of the uterine wall.

in vitro A Latin phrase that literally means "in glass," referring to some type of procedure carried out in a test tube, on a laboratory dish, or in some other artificial environment. *See also* **in vivo.**

in vitro fertilization (IVF) An artificial method of reproduction in which male and female cells are joined to each other outside the human body.

in vivo A Latin phrase that means "in life," referring to a procedure that occurs within a living organism, as within a laboratory animal or a human. *See also* **in vitro.**

informed consent Permission granted by a person to participate in a research decision, based on that person's understanding of the potential risks and benefits associated with his or her participation.

inner cell mass (ICM) A cluster of cells within the blastocyst that eventually gives rise to the embryonic disk of the embryo and, ultimately, to the fetus.

long-term self-renewal The process by which stem cells replicate themselves without differentiating (that is, they remain stem cells) over long periods of times, ranging from a few months to many years.

mesenchymal stem cell According to one definition, "a multipotent cell found in embryonic connective tissue and, much more rarely, in adult bone marrow and connective tissue; capable of differentiating into bone, cartilage, and fat cells" (Dirckx, at http://www.hpisum.com/perspectives/issue50/update.pdf). However, one authority in the field of stem cell research has reviewed the literature and concluded that "there is no accepted definition of a mesenchymal stem cell, not even an operational one" (Horwitz, at http://www.celltherapy.org/committees/Committees/Mesenchymal/mcis01.htm).

mesoderm The middle layer of the three layers of cells that make up the embryo. The mesoderm eventually gives rise to cells that make up connective tissue, muscles, bones, blood, the genital system, and some glands.

morula A ball of cells with the appearance of a mulberry (hence, the name) that forms three to four days after fertilization. The morula consists of 16, 32, or 64 cells.

multipotency The ability of a stem cell to differentiate into more than one type of tissue cell, although all of the cells into which it differentiates are of the same tissue type. For example, a blood stem cell may differentiate into any one of a variety of white blood cells or red blood cells, but does not typically develop into a nerve, muscle, skin, or nonblood type of cell.

murine Pertaining to mice or rats.

neural stem cell A type of stem cell found in human nerve (neural) tissue that develops into various kinds of nerve cells.

neuron A nerve cell.
niche A matrix of tissue cells and molecules in which one or more stem cells is embedded and which controls self-renewal and prevents the differentiation of those stem cells.
oligopotent progenitor cell A progenitor cell with the capability of differentiating into more than one, but only a limited number, of different kinds of cells.
oocyte (oöcyte) A female gamete, or egg cell.
ovum *See* **oocyte.**
parthenogenesis The development of an unfertilized egg into a mature individual. The process occurs naturally in some animals and has been produced artificially in other animals. The process has been proposed as a method for producing embryonic stem cells without the intermediary step of fertilizing an egg.
plasticity The ability of a stem cell to grow and differentiate into a different kind of cell.
pluripotency The ability of a stem cell to develop into many different kinds of cells. The term often refers to the ability of a stem cell to differentiate into all kinds of cells found in the postimplantation embryo, fetus, or developed organism, but not in the trophoblast or placenta (so-called extra-embryonic entities).
preembryo A term sometimes used to describe the earliest stages of life, ranging from the fertilized egg to any larger entity in which cells have not yet begun to differentiate. The term is the subject of a great deal of dispute, with many experts in the field of embryology suggesting that it has no scientific meaning and is used only for political purposes.
preimplantation genetic diagnosis and screening (PGD) A set of procedures in which embryos that have been created by in vitro fertilization are tested for certain genetic traits so as to determine which of those embryos is to be implanted.
primitive streak A band of cells that develops about 14 days after fertilization along the longitudinal axis of the body that later becomes the fetal spinal cord.
progenitor cell A cell present in fetal or adult tissue that, like a stem cell, can differentiate into another kind of specialized cell. Unlike a stem cell, however, it is unable to continually renew itself by repeated cell division.
proliferation The multiplication of a single cell or small group of cells into a large population of identical cells through the process of cell division.
regeneration In medicine, the process by which an organism regrows tissue, organs, or some other body part.
regenerative medicine A field of medicine in which stem cells are introduced into a person's body and induced to differentiate into some specific type of cell tissue in order to repair or replace damaged tissue.

reparative medicine *See* **regenerative medicine.**

reproductive cloning (1) A process by which a complete new organism is created by somatic cell nuclear transfer (SCNT) beginning with a single body cell of another organism, to which the new organism is genetically identical; (2) cloning of an embryo for transplantation into a uterus in order to produce a mature organism that is genetically identical to the nuclear donor.

somatic cell Any cell that is not a germ cell, that is, a sperm or egg cell.

somatic cell nuclear transfer (SCNT) The process by which the diploid nucleus of a somatic cell is transplanted into an unfertilized oocyte from which the nucleus has been removed. When this chimeric cell begins to divide, it produces totipotent stem cells that are genetically identical to the donor of the diploid nucleus.

somatic stem cell *See* **adult stem cell.**

stem cell A cell that is capable of dividing over some indefinite period of time and differentiating to produce one or more kinds of specialized cells.

stem cell niche *See* **niche.**

"stemness" A term that is often applied to a stage in a cell's life during which it has the properties of a stem cell (ability to divide over many generations and to eventually differentiate into a specialized cell) that can be characterized by certain biological and chemical characteristics of the cell.

teratogeny The experimental study of teratomas.

teratoma A tumor that contains tissues from all three embryonic germ layers—endoderm, ectoderm, and mesoderm—most commonly found in the gonads (ovaries and testes).

therapeutic cloning The use of somatic cell nuclear transfer (SCNT) to produce an embryo that is allowed to develop to the blastocyst stage, at which point it is sacrificed for the purpose of harvesting embryonic stem cells contained within it.

totipotency The capacity of a stem cell to differentiate into any one of the 210 different types of cells normally found in the human body, or into all types of the cells found in some other organism.

transdifferentiation The process by which an adult stem cell from one kind of tissue differentiates into a cell of a different type of tissue.

trophoblast The outer layer of cells in the blastocyst that attaches to the inner wall of the uterus during implantation of the embryo, eventually becoming the placenta through which food passes to the embryo and, later, the fetus.

unipotency The ability of a stem cell to differentiate into a single type of mature somatic cell.

zygote Technically, a diploid cell formed by the fusion of two haploid cells during sexual reproduction. More commonly, a fertilized egg that has been formed by the fusion of a sperm cell and an egg cell.

PART II

GUIDE TO FURTHER RESEARCH

CHAPTER 6

HOW TO RESEARCH STEM CELL ISSUES

The subject of stem cell research has been in the forefront of national and international news for the past decade. This chapter suggests a number of ways in which researchers can learn more about the scientific background, ethical issues, and legal status of stem cell research and the positions that are being argued with respect to those issues.

This chapter provides suggestions about the use of print and electronic resources generally available to researchers. In addition, suggestions are offered for the somewhat specialized area of legal issues relating to stem cell research.

PRINT SOURCES

For many centuries, the primary source of information on any topic has been the library. Libraries are usually buildings that hold materials known as *bibliographic resources*, books, magazines, newspapers, and other periodicals and documents, as well as audiovisual materials and other sources of information. General libraries tend to vary in size from small local facilities with only a few thousand books and periodicals to mammoth collections like the Library of Congress in Washington, D.C.; the Bibliothèque Nationale de France, in Paris; and the British Library, in London, each of which holds millions of individual items. You can find a list of the world's major general national libraries online at "National Libraries of the World" (http://www.ifla.org/VI/2/p2/national-libraries.htm).

Some libraries specialize in specific topics, ranging from business and education to medical research and the health sciences. Examples of specialized libraries are the Monroe C. Gutman Library (education) at Harvard University; the Gulf Coast Environmental Library in Beaumont, Texas; the

Jonsson Library of Government Documents at Stanford University; the Cornell Law Library, in Ithaca, New York; and the health sciences libraries maintained by many universities.

Specialized libraries often have information on a topic that is not available at most general libraries. At one time, that fact was not very helpful to someone who would have to travel to Berkeley, Princeton, or some other distant location to obtain the information he or she needed. Today, most libraries, both general and specialized, have online catalogs that are entirely or partially available to anyone with access to a computer. These online catalogs often allow a researcher to locate a needed item, an item that can then be ordered through a local library by means of interlibrary loan. Further information about general and specialized libraries and about the use of interlibrary loan services can be obtained from your local school or community librarian. One particularly useful reference in locating health sciences libraries is the web site "Medical/Health Sciences Libraries on the Web" (http://www.lib.uiowa.edu/hardin/hslibs.html). The site lists specialized libraries in virtually every state, with direct electronic connections to most of those libraries.

LIBRARY CATALOGS

The key to accessing the vast resources of any library is the card catalog. At one time, a card catalog consisted exclusively of a collection of cards stored in wooden cabinets and arranged by title, author, and subject of all the books and other materials owned by a library. Today, most libraries also have an electronic card catalog in which that information exists in electronic files that can be accessed through computers. Some libraries have eliminated the older, physical form of their card catalog, making it possible to access their collections only through the electronic card catalog. The electronic card catalog has the advantage of being available to researchers from virtually any location, compared to the traditional physical card catalog, which can be accessed only at the library itself.

Whether one searches a library's resources by means of its physical card catalog or its electronic equivalent, that search may take any one of a number of forms. One may, for example, search for the title of a publication, by the author, by subject matter, by certain key words, by publication date, or by some other criterion. While physical catalogs tend to use only the first three of these criteria, electronic catalogs often provide researchers with a wider range of options, options that can be explored by means of *advanced searches.* Advanced searches allow one to search for various combinations of words and numbers, combining some terms, and requiring that others be ignored. For example, one may wish to locate books that have been written on the subject of stem cells only between the years 2000 and 2005, only in English, and only by an author with the last name of Black. An advanced search allows these conditions to be used in looking for items in a catalog.

Advanced searches are very helpful when an initial search produces too many results. If one looks only for the subject *stem cells* in a catalog, for example, one may find hundreds or thousands of entries. For example, a search for that term in the online Summit search engine, which is used by all academic libraries in the state of Oregon returns a total of 131 items. While not impossible, examining all the titles in that list would be time-consuming, especially if one knows in advance that he or she is interested in only one aspect of the subject. For example, if the topic of interest were really legal issues involving stem cells, an advanced search should be used that combined these terms, in the form *"stem cells" and "law"*, *"stem cells" and "legal issues"*, or some similar choice of search terms. Combinations such as these return only 19 items and two items, respectively, in the Summit system—far more manageable lists.

One of the key tools used in advanced searches is the Boolean operator. A Boolean operator is a term that tells a computer how it should treat the term(s) surrounding another term. The three most common Boolean operators are AND, OR, and NOT. If a computer sees two words or phrases connected by an AND, it understands that it should look only for materials in which *both* words or phrases occur. If the computer sees two words connected by an OR, it knows that it should search for any document that contains one word or phrase *or* the other, but not necessarily both. If the computer encounters two words or phrases connected by a NOT, it understands that it should ignore the specific category that *follows* the NOT when searching for the general category that precedes the NOT.

An important skill in searching for materials in either a library or on the Internet is to remember that documents are not always identified by a computer in the same terms in which a researcher is thinking of them. For example, a researcher may be interested in tracking down all books in a library on the subject of *stem cell research*. If he or she types that term into the library's catalog, a few hundred or a few thousand items may show up. What the researcher may not know is that many more items of interest may exist in the library's holdings. Other authors and/or librarians may use other terms to describe the same idea as expressed by *stem cell research*. Some terms that might also refer to articles or books on *stem cell research* are *stem cells*, *embryonic research*, and *fetal research*. These terms do not mean exactly the same thing as *stem cell research*, but they are close enough to serve as search terms. But even these terms do not exhaust the possible range of identifiers that will produce materials of value to the researcher. Some other words and phrases one might try include the following:

- adult stem cells
- cloning
- embryonic stem cells
- gene therapy
- human embryology
- in vitro fertilization
- religion and medicine
- stem cell transplantation

In fact, a library's search engine often suggests terms related to the one for which you searched. In a search on the Summit engine for "stem cells" and "legal issues," for example, the search engine suggested also looking for terms such as biotechnology, biotechnology—patents, biotechnology industries, pharmaceutical industry, drugs, research and development partnership, patent laws and legislation, technological innovations—economic aspects, genetic engineering, human cloning, bioethics, and intellectual property—valuation, among other possible terms.

The same approach is necessary, of course, in searching for more specialized areas of stem cell research. In looking for materials on the subject of ethical issues related to stem cell research, one should search not only under that term, but also under a variety of other terms, such as:

- bioethics
- ethical issues in medicine
- stem cell research—ethical issues
- stem cell research—pros and cons

One of the keys to success in finding all or most of the materials in a library on some given topic is to imagine as many different words and phrases as possible by which that topic might be identified.

Scholarly Articles

One area in which libraries continue to have an important advantage over Internet searching is in the use of scholarly articles. Anyone interested in stem cell issues will want to examine articles published in all kinds of periodicals, ranging from general interest newspapers, such as *The New York Times* and the *Washington Post*, to more specialized journals, such as *The American Journal of Bioethics*, *Stem Cell Business News*, *Stem Cell Research News*, and *Stem Cells: The International Journal of Cell Differentiation and Pro-*

liferation. Many libraries—especially academic and specialized libraries—subscribe to a wide variety of periodicals such as these, both general and specialized in character. Anyone who visits the library has access to these periodicals free of charge.

Those periodicals are, in almost all cases, also available online. Periodicals differ in their policies on accessing past articles. In some cases, they allow access free of charge to all articles for all users of the Internet. In other cases, they make some articles available at no cost, while a charge is assessed for accessing other articles. In the majority of cases, scholarly journals restrict access to articles to members of some particular professional society (such as the American Society for Cell Biology) and to readers who are willing to pay for an article. Nonmembers may purchase the right to read an article for a fee, which varies from periodical to periodical. In many cases, the fee is substantial, ranging from $10 to $35 per article. The general researcher will seldom be able to afford the purchase of every article found on the Internet on restricted sites. His or her choice, then, is to try locating that article in a local library (usually an academic library), requesting a copy of the article through interlibrary loan, or purchasing the article online from the journal publisher.

INTERNET SOURCES

Today, the resources of libraries have been greatly enhanced by the Internet and its cousin, the World Wide Web. The Internet is a vast collection of networks, each containing very large amounts of information, generally accessible to almost any individual computer. The Internet was first created for the U.S. military in 1969 and has since expanded to include networks for every imaginable use for general users in all parts of the world.

WEB SITES

On the Internet, data are stored in web sites, locations where information about some specific topic is to be found. That information may range from the very specific to the very general. In the case of stem cell research, for example, one can find web sites that focus on topics as specific as a particular research study on stem cells (such as "Stem Cells Promise Liver Repair" at http://news.bbc.co.uk/1/hi/sci/tech/841932.stm) or as general as the overall subject of stem cell research itself (such as "Stem Cells in the Spotlight" at http://gslc.genetics.utah.edu/units/stemcells).

A good place to begin in researching a topic such as stem cell research is with a web site whose primary or exclusive focus is on this subject. Some examples are:

- Center for Human Progress, "Stem Cell Research" (http://www.americanprogress.org/site/pp.asp?c=biJRJ8OVF&b=669211)
- Discovery Media, "Stem Cell Research Links" (http://www.stem-cell-research-links.com)
- HavenWorks.com, "Stem Cell.news" (http://www.havenworks.com/health/stem-cell)
- Questia, "Stem Cell Research" (http://www.questia.com/Index.jsp?CRID=stem_cell_research&OFFID=se1&KEY=stem_cell_research)
- Wikipedia, "Stem Cell" (http://en.wikipedia.org/wiki/Stem_cell)

Additional web sites on stem cell research can be found in Chapter 7 of this book.

One benefit of general purpose web sites is that they often provide links (connections) to other web sites with information on similar or related topics. The Wikipedia web site listed above, for example, provides links to the following web sites:

- "Regenerative Medicine," a special supplement of the *Proceedings of the National Academy of Science*
- Judith A. Johnson and Erin Williams, *Stem Cell Research*, a publication of the Congressional Research Service
- WebMD's "Stem Cells Q & A"
- "Stem Cells: Policies and Players," a Genome New Network web page
- "Stem cells and Genesis," a Christian perspective on stem cells that opposes embryonic stem cell research
- Cord-Blood.org, a web site with information on cord blood banking

Searching through web sites on the Internet involves three fundamental problems of which the researcher should always be wary: complexity, accuracy, and evanescence (instability). Internet web sites are related to each other in a complex, weblike fashion (hence the name World Wide *Web*), and not in a linear fashion, like the chapters in a book. When one goes looking through the Internet ("surfing the net"), one quickly heads off in dozens of different directions, often crossing and crisscrossing pathways and web sites. It is easy to get lost and forget how one arrived at a particular web site or how to get back to the beginning of a search string. One technique for keeping track of data is to print out every page that may seem to have some significance to one's research. That information may later prove to be of little or no value. But if it does turn out to be important, the researcher does not have to worry about finding it again at some time in the future. One can also mark the page containing the information as a "Favorite," using the toolbar at the top of the search engine's home page, allowing one to return to that page if and when it is needed at a later time.

The second inherent problem with the Internet is the accuracy of web sites. Anyone can create his or her own web site with any kind of information at all on it. The information does not have to be true or accurate, and no outside monitor exists to tell a researcher whether the information is reliable or not. In searching for information on stem cell research, for example, one web site could report that the first therapeutic use for stem cells will appear within the next 12 months, will provide cures for all known diseases, and will lengthen human life span to 125 years. There is no way of knowing which of these statements is (or are) actually factual. As a result, researchers must constantly be even more careful than they are with print materials (which are, at least, usually edited and fact checked) as to the accuracy of information found on the Internet. They can, for example, check other web sites or print materials on the same topic to verify that information given as facts is really true. They can also look carefully at web sites themselves to see if they appear to have some special argument about stem cell research in their presentation.

Bias, and the inaccuracies it may include, is especially likely to occur in web sites on controversial issues. When people feel very strongly about a topic, they may accidentally or intentionally provide information that is incomplete, slanted, or simply wrong. Researching a topic such as stem cell research, therefore, requires a degree of caution and a willingness to double-check information more often than when conducting other forms of research.

The third problem in conducting online research is evanescence, or the tendency of web sites to disappear over time. Nothing is likely to be as frustrating for a researcher as to find a reference to a web site with what looks to be just the right information, only to receive the message "Web site not found." The web site has, usually for unknown reasons, been deleted, and is no longer available on the Internet.

However, it may not really be gone forever. Sometimes the web site's address has simply been changed, a possibility that the search engine may suggest by providing possible alternative leads to the site. Or, the web site may have been *cached*, that is, set aside in a "hidden" location in the computer's memory. The web site may then be accessed by asking the search engine to look into its "hidden memory" and pulling up the desired page. Many "dead sites" have been stored on the Internet Archive (http://www.archive.org). A word of caution: Not every site is stored here, and users need to know the exact URL to access a site. Also, most sites archived have been stripped of all their image files.

WEB INDEXES

One kind of web site—a *web index*—is of special interest to researchers. A web index, as the name suggests, is a web site that is organized like an outline, arranged according to subject matter and then divided into related subtopics. Web indexes are so-called because they are, in a sense, similar to the index in a book.

One of the oldest and most popular web indexes is Yahoo! (http://www.yahoo.com). Two other very popular web indexes are LookSmart (http://search.looksmart.com) and About.com (http://about.com). Yahoo!'s home page provides two ways of searching a web site's content. First, one can simply type a word, phrase, or set of words into a search box. Yahoo! will then act like a search engine on its own site, looking for any and all web pages that match the required term(s).

Second, one can select one of the major topics listed on Yahoo!'s main page and work his or her way through increasingly more specific subtopics. For example, to find web sites on stem cell research, one would go first to one of the major categories listed, such as *Science* or *Health.* The next step is to look into each of these major categories to find subcategories that are likely to contain entries on the topic, *Diseases and Conditions,* or *Health Sciences,* under *Health,* for example, or *Biology* or *Medicine,* under *Science.* Stem cell research happens to be a subject that appears in more than one category, so the researcher will have to look in all of those categories to find all the web pages related to this term. Using the "only within this category" button at the top of the page will help with this search.

One advantage of web indexes is that they tend to be selective. That is, rather than searching for any or all web sites that feature the subject of stem cell research, for example, they try to determine the potential usefulness of such sites. In this way, researchers are less likely to have to wade through dozens of web sites with outdated, strongly biased, inaccurate, or otherwise less useful information.

One problem with some web indexes (as with may web pages and some search engines) is the prevalence of "pop-up" advertising. "Pop-ups" are windows that appear automatically on opening a page. Sometimes they are related to the topic, sometimes they simply advertise general topics ("Look for Your High School Sweetheart"). Pop-ups are an important source of revenue for Internet companies and are likely to become more common in the future. They are usually a headache for researchers, for whom they are nothing other than a time-consuming distraction. One way to avoid pop-ups is to install software that recognizes and hides pop-ups as soon as they appear. Another way is to avoid certain web indexes (such as About.com) in which pop-ups tend to appear more commonly than in other sites.

SEARCH ENGINES

Other tools that can be used in Internet research are search engines. Search engines are systems by which one can sift through the millions of web sites available online in order to find those that may contain information on some topic of interest. Search engines are amazing technological tools that accomplish this objective in a fraction of a second and then present the researcher with the names of web sites that are likely to be of interest. The difference between search engines and web indexes is not always clear, nor is it usually important that such distinctions be made.

Some of the most popular search engines now available include Google, Yahoo!, Dogpile, Ask Jeeves, AllTheWeb, HotBot, Teoma, and LookSmart. (For an exhaustive review of search engines and related topics, see SearchEngineWatch at http://searchenginewatch.com or SearchEngineJournal at http://www.searchenginejournal.com.) By far the most widely used of these engines is Google, which claims to search out more than eight billion web sites.

Learning to use a search engine is similar to learning other skills: The longer one practices, the more one learns about the process and the better one becomes at it. The easiest approach is simply to type in a word or phrase in which one is interested and press "Enter." The problem is that "easy" in searching is often not "efficient." A search engine is likely to return the names of many web sites that have little or nothing to do with the topic of interest. For example, asking a search engine to look for *stem cell research* may produce a number of web sites that discuss books about stem cell research, companies engaged in stem cell research or the manufacture of equipment and materials for stem cell research, blogs (online personal journals) about stem cell research, or some other topic related to stem cell research. While these topics may be of interest to a researcher, they may also be too specialized or to far afield from the researcher's main field of interest in stem cell research.

One way to avoid having too many unrelated web sites appear during a search is to be as specific as possible. If one wished to obtain information on the current status of the research on adult stem cells, for example, it would not be very efficient to start searching just for *stem cell research*. Even though that search probably would turn up web sites that discuss adult stem cell research, it would also produce many pages that have nothing to do with that specific topic. Instead, it would be more efficient to ask for some combination of terms, such as *stem cell research*, *adult stem cells*, *current research*, and *2005*, for example. Or, an even more efficient approach might be to ask for the specific topic in which one is interested, such as "adult stem cell research progress in 2005." Notice that words within quotation marks will be treated by the Search engine as unitary terms. A web site that uses the phrase *stem cells* rather than *stem cell* will not be listed by the search engine. Nor will there be a web site in which the term *adult stem cell* only is listed or one where the term *stem cell* is misspelled.

Trial-and-error and "practice, practice, practice" are two good ways to improve one's skills on Internet searching. Formal instruction is often very helpful also. One can learn a great deal from a brother or sister, a teacher, or a friend who has experience working on the Internet. Instruction is also available on the Internet itself. For example, three web sites that provide tutorials on Internet searching are Learn the Net at http://www.learnthenet.com/english/index.html, Silwood Cyber Centre at http://www.silwoodonline.org.uk/cybercentre/learning.htm, and the University of California-Berkeley Internet tutorial at http://www.lib.berkeley.edu/TeachingLib/Guides/Internet/FindInfo.html.

Another type of search engine that is often of value is the metasearch program. Metasearch engines (also known as *metacrawlers*) hunt through other search engines, collecting web sites in each of those search engines they believe to be the best matches of a researcher's search terms. Among the most popular metasearch engines now available are Dogpile, Vivisimo, Kartoo, Mamma, and SurfWax.

STEM CELL RESEARCH WEB SITES

Some of the most useful information about stem cell research issues is to be found in web sites that are wholly devoted to that subject. Those web sites may range in size from a single page to one with dozens or even hundreds of pages. The goals of such sites also range from providing a brief, general overview of the subject to covering as many aspects of the issue, such as scientific, technical, economic, social, and political aspects, as possible. An example of a general information web site on stem cell research is the one maintained by the National Institutes of Health, "Stem Cell Information" at http://stemcells.nih.gov/index.asp.

Specialized web sites also differ from each other in the position they take on various aspects of stem cell research. Some sites attempt to provide information in a neutral manner, offering factual information alone and allowing readers to make up their own minds about the subject of stem cell research. An example of this kind of web site is the excellent site on stem cells on the Genome News Network web page at http://www.genomenewsnetwork.org/categories/index/stemcells.php. Other web sites exist to promote a particular point of view about stem cell research, either attempting to encourage its support and development or opposing its use in all, or at least some, applications. Two examples of such web sites are (pro–stem cell research) "Stem Cell Action" (http://stemcellaction.org) and (anti–stem cell research) "Stem-cell Research and the Catholic Church" (http://www.americancatholic.org/News/StemCell/default.asp). As with any resource, researchers must always be aware of any biases present in a web site and determine the accuracy of the information from the web site, given such bias.

LEGAL RESEARCH

In one regard, much of the debate over stem cell research is essentially a debate over legal issues. What kinds of stem cell research should be allowed? To what extent should the federal government and/or state government monitor and control stem cell research? Should governmental agencies be allowed to provide funds for stem cell research and, if so, for what kinds of research? Questions such as these are all answered eventually by the passage of laws at the local, state, or federal level or by the imposition of rules and regulations by executive and/or regulatory bodies. The search for legal information is of special interest and importance, therefore, for researchers interested in stem cell–related issues.

FINDING LAWS AND REGULATIONS

The best places to begin a search for laws and regulations in regards to stem cell research are web sites that specialize in such information. These web sites are maintained both by relatively disinterested governmental agencies and by special interest groups and individuals who may or may not have some stake in supporting or opposing stem cell research. Some examples of these web sites including the following:

- "Background and Legal Issues Related to Stem Cell Research" (http://fpc.state.gov/documents/organization/11274.pdf)
- Genome News Network, "Stem Cells: Policies and Players" (http://www.genomenewsnetwork.org/resources/policiesandplayers)
- Grass Roots Connection, "GRC State Stem Cell Research News and Alerts" (http://grassrootsconnection.com/state_stem_cell_resources.htm)
- HumGen, "GenBiblio" (http://www.humgen.umontreal.ca/int/GB.cfm)
- National Institutes of Health, "Policy & Guidelines" (http://stemcells.nih.gov/policy/guidelines.asp)
- University of Minnesota Medical School, "Stem Cell Policy: World Stem Cell Map" (http://mbbnet.umn.edu/scmap.html)

Thus far, relatively few laws have been passed on a national level dealing with stem cell research. That situation may, of course, change in the future.

One of the best guides to existing laws on stem cell research, or on any other subject, is the Library of Congress's Thomas web site (http://thomas.loc.gov), which provides complete information on all bills introduced into both the U.S. House of Representatives and the U.S. Senate, along with the ultimate fate of those bills. The *Congressional Record*, dating back to 1989, is also available on the web site.

Another indispensable web site is FindLaw (http://lp.findlaw.com), a web site designed for both legal professionals and laypersons. FindLaw provides information on virtually every imaginable area of law, including not only federal law, but also legal codes for all of the individual states. Since most states have their own laws and regulations dealing with stem cell research that cannot be found on Thomas, FindLaw is the essential companion to the Library of Congress's site in carrying out a complete search of laws and regulations on stem cell research.

FINDING COURT DECISIONS

Passing laws and adopting regulations are only one step in the regulation of stem cell research in the United States and other nations. Ultimately, most laws are tested in court to produce a (usually) final decision as to exactly what those laws and regulations allow and prohibit.

Locating court decisions on the Internet is somewhat similar to the process of finding laws and regulations described in the preceding section. Web sites that contain information on laws and regulations often include additional information on the way those laws and regulations have been interpreted by the courts.

For researchers with little background in legal matters, an excellent tutorial is available on the Internet. The tutorial is called Legal Research FAQ (*FAQ* stands for "frequently asked questions") and was authored by attorney Mark Eckenwiler in 1996. The tutorial provides a very readable and detailed explanation of the way court decisions are identified and how they can be located online and in print resources. The tutorial can be accessed by a number of pathways, one of which is http://www.faws.org/faws/law/research. A number of other web sites provide suggestions for searches in specific libraries or other sources. For example, the Law Library of Congress maintains a site of this kind at http://www.loc.gov/law/public/law-faq.html. An excellent overview on using the Internet for all kinds of legal research has also been made available by Lyonette Louis-Jacques, librarian and lecturer in law at University of Chicago Law School. That overview, "Legal Research Using the Internet," is available online at http://www.lib.uchicago.edu/~llou/mpoctalk.html.

CHAPTER 7

ANNOTATED BIBLIOGRAPHY

Stem cell research has been a topic of interest to the scientific community and to the general public at large for only a relatively short period of time, probably less than a decade. Still, a number of reports, books, articles, Internet web pages, and other documents have already been produced on this topic. This chapter lists a number of these documents of interest to anyone researching the topic of stem cell research. The chapter is divided into five sections dealing with general introductions to the subject (including both scientific and nonscientific topics), the scientific and technological background of stem cell research, its history as a social issue, laws and regulations relating to stem cell research, and moral and ethical arguments that have been presented both for and against the continuation of stem cell research. Within each of these sections, documents have been arranged according to format: books, journals, articles, reports, and web documents.

Some overlap is to be expected within this system of classification. That is, some documents present a comprehensive discussion of stem cell research, attempting to provide a review of scientific and technical topics, a history of the subject, and arguments both for and against the use of stem cell research. In such cases, the document is listed in the category that describes its greatest emphasis. Also, some documents may appear in more than one format. Reports on stem cell research are often reissued, for example, in book format or on an Internet web site. In such cases, the document is listed in what is probably its most accessible format, with mention of other formats in which it is also available.

The abbreviation SCR is frequently used in this chapter to stand for *stem cell research*.

GENERAL INTRODUCTION

BOOKS

Allman, Toney. *Stem Cells.* Farmington Hills, Mich.: Lucent Books, 2005. A general introduction to stem cell research and related issues.

Blazer, Shraga, and Etan Z. Zimmer, eds. *The Embryo: Scientific Discovery and Medical Ethics.* New York: Karger, 2005. A collection of papers that discuss various issues on embryonic life, including the beginning of life; embryonic stem cells; societal, ethical and religious views on genetic intervention in humans; fetal surgical and pharmacological intervention; fetal imaging and monitoring; and law and justice.

Committee on the Biological and Biomedical Applications of Stem Cell Research, Board on Life Sciences National Research Council. *Stem Cells and the Future of Regenerative Medicine.* Washington, D.C.: National Academy Press, 2002. A report of a workshop held on June 22, 2001, at which many leading researchers in the field of stem cells discussed their work, and a group of philosophers, ethicists, and legal scholars commented on the political, social, moral, and other implications of this research. The book is available online at http://www.nap.edu/books/0309076307/html.

Espejo, Roman, ed. *Human Embryo Experimentation.* Farmington Hills, Mich.: Greenhaven Press, 2002. An anthology of articles designed for readers aged nine to 12 years on the science and ethical implications of research on the human embryo.

Panno, Joseph. *Stem Cell Research: Medical Applications and Ethical Controversy.* New York: Facts On File, 2004. Although written for young adults, this book should be of interest to a broad range of adult readers also. It provides a clear review of scientific advances in stem cell research as well as legal and ethical issues posed by this research.

Prentice, David A., and Michael A. Palladino. *Stem Cells and Cloning.* Menlo Park, Calif.: Benjamin Cummings, 2002. The authors describe and discuss two of the most important new biological procedures developed in recent history, cloning and stem cell research, along with a commentary on the ethical, legal, and other issues raised by research in these fields.

Shostak, Stanley. *Becoming Immortal: Combining Cloning and Stem-Cell Therapy.* New York: New York University Press, 2002. Speculations on the possibility that scientists may be able to overcome the natural processes of aging and death and the implications of such an advance for the nature of human life, commerce, politics, human dignity, and technology.

Tesar, Jenny E. *Stem Cells.* Farmington Hills, Mich.: Blackbirch Press, 2003. A general introduction to the subject of stem cell research for readers aged nine to 12 years. One reviewer for the journal *Booklist* called the books in this series (Science on the Edge) "[e]xemplary in their balanced, easy-to-grasp coverage of complex issues."

West, Michael. *The Immortal Cell: How Stem Cell Biotechnology Can Conquer Cancer and Extend Life.* New York: Doubleday, 2003. The author is the founder of Advanced Cell Technology, the only U.S. company known to be conducting human cloning research for medical purposes. The book is a general explanation as to how stem cells work and how they can be used in medical therapeutics.

ARTICLES

Bangsbøll S., et al. "Patients' Attitudes Towards Donation of Surplus Cryopreserved Embryos for Treatment or Research," *Human Reproduction,* vol. 19, no. 10, October 2004, pp. 2,415–2,419. Report of a study of 284 couples who had produced fertilized eggs for possible use in in vitro fertilization, but whose eggs had never been used. The researchers asked what use the couples were willing to permit for those eggs, including donation to other infertile couples and use in research, such as stem cell research. More than half of the couples agreed that they would permit the fertilized eggs to be used for the latter purpose, but only a quarter agreed to the former use.

Birmingham, Karen. "Europe Fragmented over Embryonic Stem Cell Research," *Journal of Clinical Investigation,* vol. 112, no. 4, August 15, 2003, p. 458. Regulations imposed on stem cell research by various European countries and by the European community itself vary widely, resulting in a condition that one researcher interviewed for the article calls "an absolute disaster."

Cogle, Christopher, R., et al. "An Overview of Stem Cell Research and Regulatory Issues," *Mayo Clinic Proceedings,* vol. 78, no. 8, August 2003, pp. 993–1,003. Also available online. URL: http://www.mayoclinicproceedings.com/inside.asp?AID=401&UID=#bib120. The article discusses the science of stem cell research and its potential medical applications in detail and briefly alludes to ethical issues and regulatory restrictions in the United States.

Cohen, Cynthia B. "Stem Cell Research in the U.S. after the President's Speech of August 2001," *Kennedy Institute of Ethics Journal,* vol. 14, no. 1, March 2004, pp. 97–114. A general overview of the conditions on stem cell research outlined during President George W. Bush's speech and the impact these regulations are likely to make on the practice of stem cell research in the United States.

Cyranoski, David. "Stem-cell Research: Crunch Time for Korea's Cloners," *Nature,* vol. 429, no. 6987, pp. 12–14. Also available online. URL:

http://www.nature.com/news/2004/040503/pf/429012a_pf.html. A review of the scientific accomplishments of Hwang Woo Suk's stem cell research team, the response his work has produced in South Korea, and some possible ethical and legal problems facing future research of this kind in South Korea.

Defrancesco, Laura. "Stem Cell Researchers Take on Parkinson's," *The Scientist*, vol. 15, no. 11, May 28, 2000, p. 1ff. A report on the use of stem cells to increase the concentration of dopamine in the brain, with the subsequent promise of treating diseases like Alzheimer's and Parkinson's. There is some concern that political factors may be hindering the rate at which such research can be conducted.

Fox, Cynthia. "Why Stem Cells Will Transform Medicine," *Fortune*, vol. 143, no. 12, June 11, 2001, pp. 184–195. A general description of the process by which stem cells are cultured, extracted, and used, and the implication for this research for bringing about substantial changes in the medical profession. The article also discusses the business implication of this change.

Goldstein, Andrew. "The Great Cell Debate," *Time*, vol. 158, no. 3, July 23, 2001, pp. 24–25. A general introduction to the science and ethics surrounding stem cell research.

Goodman, Laurie. "States Step in for Stem Cell Research," *Journal of Clinical Investigation*, vol. 113, no. 10, May 15, 2004, pp. 1,376–1,377. A review of the steps that some states have taken to support the funding of stem cell research, given the reluctance of the federal government to do so.

Hines, Pamela J., Beverly A. Purnell, and Jean Marx. "Stem Cells Branch Out," *Science*, vol. 287, no. 5457, February 25, 2000, p. 1417. An introduction to a special edition of *Science* magazine that discusses the scientific and technological status of stem cell research as well as some of the ethical, legal, and political issues involved in such research. A good general overview of the state of the art as of early 2000.

Lacayo, Richard. "How Bush Got There," *Time*, August 20, 2001, pp. 17–23. A detailed analysis of the factors that led President George W. Bush to reach his decision about U.S. policy on stem cell research. In the same issue, there are related articles by N. Gibbs and M. Duffy, "In a 21st Century Speech on Stem Cell Funding, Bush Budges and Finds Compromise: Will It Work?" (pages 15–16) and F. Golden, "Before James Thomson Came Along, Embryonic Stem Cells Were a Researcher's Dream" (pages 27–28).

Lanza, Robert, and Nadia Rosenthal. "The Stem Cell Challenge," *Scientific American*, vol. 290, no. 6, pp. 93–99. After an introduction to the procedures involved in stem cell research, the authors explore the technical problems that must be solved before SCR can be applied to the treatment of specific medical conditions.

MacDonald, Chris. "Stem Cells: A Pluripotent Challenge," *Bioscan*, vol. 13, No. 4, Fall 2001, pp. 7–8. A discussion of the science and issues relating to embryonic stem cells.

May, Mike. "Mother Nature's Menders," *Scientific American*, vol. 283, no. 6, June 2000, pp. 56–61. An article written early in the development of human embryonic stem cell research with a somewhat glowing and naive description of its potential for medical therapies, with modest mention of the ethical issues that are likely to arise as a result of SCR.

Ready, Tinker. "Private Donors Breathe New Life into U.S. Stem Cell Research," *Nature Medicine*, vol. 10, no. 4, April 2004, p. 319. A news report on the effects that private funding have had on the progress of stem cell research in the United States.

Thomas, John W. "Cell Sources and Support Programs for Stem Cell Research and Cell-Based Therapy in the USA," *Fetal Diagnosis and Therapy*, vol. 19, no. 3, May–June 2004, pp. 212–217. A review of programs by the National Heart, Lung, and Blood Institute to promote stem cell research within the guidelines provided by President George W. Bush and general proposals offered by Secretary of Health and Human Services Tommy Thompson.

REPORTS AND STATEMENTS

Alternative Sources of Pluripotent Stem Cells: A White Paper. Washington, D.C.: President's Council on Bioethics, May 2005. A collection of articles written by experts in the field of stem cell research, developed as a way of finding alternative sources of pluripotent stem cells to avoid the ethical issues created by having to use embryonic stem cells.

American Academy of Neurology and American Neurological Association. *Position Statement Regarding the Use of Embryonic and Adult Human Stem Cells in Biomedical Research. Neurology*, vol. 64, no. 10, May 24, 2005, pp. 1,679–1,680. Also available online. URL: http://www.neurology.org/cgi/content/full/64/10/1679. A statement in which members of the two groups indicated express their support for the federal funding of stem cell research that will "meet the standards of scientific and ethical oversight by external peer review."

Chapman, Audrey R., Mark S. Frankel, and Michele S. Garfinkel. *Stem Cell Research and Applications: Monitoring the Frontiers of Biomedical Research.* Washington, D.C.: American Association for the Advancement of Science and the Institute for a Civil Society, November 1999. A report that considers the scientific aspects of stem cell research and potential therapies that may be derived from that research, spiritual and religious concerns, ethical concerns, sources of stem cells and guidelines for their use, justice considerations, funding issues, and oversight and accountability.

The Church of England, The Ethical Investment Advisory Group. "Human Embryonic Stem Cell Research: On the Path to an Investment Framework," 2nd edition. London: The Church of England, April 2003. A report prepared on the impact that embryonic stem cell research may have on the church's investment policies. In spite of its name and this objective, the report provides an excellent general background on the science, technology, legal status, and ethical issues related to stem cell research.

Commission Staff Working Paper. *Report on Human Embryonic Stem Cell Research.* Brussels, Belgium: Commission of the European Communities, April 3, 2003. A report prepared within the context of the European Community's 6th Framework Programme, which called for a commitment to the "Application of knowledge and technologies in the fields of genomics and biotechnology for health," in particular, "research [that focuses on the] development and testing of new preventive and therapeutic tools, such as somatic gene and cell therapies (in particular stem cell therapies)." The report's four chapters deal with the origin and characteristics of human stem cells and their potential applications, human embryonic stem cell research, governance of human embryonic stem cell research, and socioeconomic aspects.

Department of Health, United Kingdom. *Stem Cell Research: Medical Progress with Responsibility.* London: Department of Health, June 2000. The report of a group of medical experts chaired by the Chief Medical Officer on the scientific possibilities of using stem cells to treat a variety of medical conditions, along with a discussion of legal and ethical issues involved in such research. The report concludes with a number of conclusions and recommendations regarding stem cell research in the United Kingdom.

Johnson, Judith A., and Erin Williams. *Stem Cell Research.* Washington, D.C.: Congressional Research Service, February 24, 2003; last updated on August 10, 2005. A report that discusses the legislative status of stem cell research, with background, scientific, and technological information and a summary of the arguments for and against such, research. The report is updated as new information becomes available.

Monitoring Stem Cell Research. Washington, D.C.: President's Council on Bioethics, January 2004. A report prepared for President George W. Bush on the current state of stem cell research in the United States with four major sections dealing with scientific background information on stem cell research, current federal law and policy, recent developments in ethical and policy debates, and recent developments in stem cell research and therapies. The report also has a valuable glossary and useful appendices.

WEB DOCUMENTS

ABC News. "Stem-Cell Industry, Research Evolving." Available online. URL: http://abcnews.go.com/Business/Technology/story?id=273177&

page=1. Posted November 23, 2004. Transcript of a broadcasted program, describing the business and legal aspects of stem cell research, especially with regard to the passage of Proposition 71 in California that authorized the expenditure of $3 billion in state funds for stem cell research projects.

Bush, George W. "Remarks by the President on Stem Cell Research." Available online. URL: http://www.whitehouse.gov/news/releases/2001/08/20010809-2.html. Posted on August 9, 2001. Transcript of a television address given by President Geore W. Bush in which he outlines his stand on embryonic stem cell research and announces the types of such research he will allow to proceed.

Center for Human Progress. "Stem Cell Research." Available online. URL: http://www.americanprogress.org/site/pp.asp?c=biJRJ8OVF&b=669211. Downloaded on June 12, 2005. An excellent web site that provides comprehensive information on latest developments in the field of stem cell research with particularly good coverage of the state of laws and regulations at the national and state levels. The site provides some very useful links to related sites on the same topics.

Center for Science, Technology, and Congress. "AAAS Policy Brief: Stem Cell Research." Available online. URL: http://www.aaas.org/spp/cstc/briefs/stemcells/index.shtml. Posted on August 26, 2004. A policy briefing prepared by the American Association for the Advancement of Science for the U.S. Congress outlining the background of stem cell research, governmental policy on the practice, ethical issues involved in SCR, and an update (as of 2004) on latest developments in the field.

Department of Health, United Kingdom. Available online. URL: http://www.dh.gov.uk/Home/fs/en. Downloaded on June 10, 2005. The Department of Health's web site contains reports, information, statistics, news, and publications on a number of health issues in the United Kingdom, including stem cell research, which can be accessed through the web site's search function.

Discovery Media. "Stem Cell Research Links." Available online. URL: http://www.stem-cell-research-links.com. Downloaded on June 9, 2005. A web site that claims to be "a news and information portal for stem cell research . . . including fetal and adult research from private sector, government funded, and academic research labs."

European Commission. "Stem Cells: Therapies for the Future?" Available online. URL: http://europa.eu.int/comm/research/quality-of-life/stemcells.html. Downloaded on June 9, 2005. A collection of documents collected for and presented at a conference held in Brussels, Belgium, set up by European Research Commissioner Philippe Busquin in December 2001. Videos of the 11 conference sessions are available online, as is an on-going "dialogue-on-line" on which individuals can submit information and questions about SCR.

Forum on Science Ethics and Policy. "Stem Cells: Background, Links, and News." Available online. URL: http://www.fosep.org/Stem_cells.htm. Downloaded on June 9, 2005. A good general resource with links to articles and web pages on stem cell basics, online video symposia, action by the U.S. and state of Washington legislatures, teachings from many religious faiths, statements in favor of and opposed to stem cell research, and recent articles and opinion pieces.

Government of Canada. "Stem Cell Research." Available online. URL: http://www.sciencetech.gc.ca/options/i_stem_e.shtml. Downloaded on June 8, 2005. Links to web sites that describe research in Canada on stem cells and the ethical and political debate surrounding this research.

Hall, Alan. "Awaiting the Miracles of Stem-Cell Research." Available online. URL: http://www.businessweek.com/bwdaily/dnfash/nov2000/nf20001129_858.htm. Posted on November 29, 2000. An article from Business Week Online providing a general overview of stem cell research with special attention to its possible economic impacts and a discussion of companies involved in the support of SCR.

HavenWorks.com. "Stem Cell.news." Available online. URL: http://www.havenworks.com/health/stem-cell. Downloaded on May 31, 2005. A superb web site that contains links to articles and web pages with articles and information on all aspects of SCR, including scientific and technological, legal and regulatory, and moral and ethical issues.

House of Lords, United Kingdom. "Stem Cell Research-Report." Available online. URL: http://www.parliament.the-stationery-office.co.uk/pa/ld200102/ldselect/ldstem/83/8 301.htm. Downloaded on May 28, 2005. A report prepared for the House of Lords outlining the scientific background for stem cell research, possible medical applications, ethical issues raised by SCR, and the present and possible future legal status of stem cell research in the United Kingdom.

Johnson, Alex. "Day of Reckoning for Stem Cell Research Nears," MSNBC.com. Available online. URL: http://msnbc.msn.com/id/7669266. Posted on May 13, 2005. A report prepared for broadcast, providing an excellent summary of the scientific background underlying stem cell research along with the ethical issues that have arisen in the United States about the use of SCR for medical applications. A number of useful links are provided for further information on the subject.

Medical Research Council. "Stem Cell Research." Available online. URL: http://www.mrc.ac.uk/index/public-interest/public-topical_issues/public-stem_cells.htm. Downloaded on June 1, 2005. A web site that provides a comprehensive review of the status of stem cell research in the United

Kingdom, listing the MRC's research plans, latest news on stem cell research, and an overview of the UK Stem Cell Initiative.

Online News Hour. "Stem Cell Research," PBS Online. Available online. URL: http://www.pbs.org/newshour/health/stemcells.html. Downloaded on June 2, 2005. A list of programs broadcast on PBS (Public Broadcasting Service) dealing with stem cell research, with links to transcripts and audio recordings of all those programs. Included are extended interviews with a number of experts in the field, latest news on advances in SCR, reports on research in the United States and other nations around the world, and discussions of the political and ethical issues surrounding SCR.

Questia. "Stem Cell Research." Available online. URL: http://www.questia.com/Index.jsp?CRID=stem_cell_research&OFFID=se1&KEY=stem_cell_research. Downloaded on May 27, 2005. Questia is a subscriber service that claims to be the world's largest online library service. The web site accesses books, articles, encyclopedias, reports, and other sources of information.

Radda, Sir George. "Stem Cell Research." Available online. URL: http://www.mrc.ac.uk/index/public-interest/public-topical_issues/public-stem_cells/public-stem_cells_house_article.htm. Downloaded on May 5, 2005. Reprint of an article that originally appeared in *The House Magazine* on April 21, 2003 regarding the status of stem cell research in the United Kingdom.

Smith, Gina. "Therapeutic Cloning, and Stem Cell Research," Naked Scientist. Available online. URL: http://nakedscientists.com/HTML/Columnists/ginasmithcolumn1.htm. Downloaded on May 15, 2005. Provides a general overview of the technology of stem cell research, its possible applications, and issues surrounding its use. The Naked Scientist web site also has other articles relating to stem cell research.

"Stem Cell." Available online. URL: http://stemcellshq.com. Downloaded on June 8, 2005. A web site that contains links to dozens of specialized topics in the field of stem cell research. A highly commercialized site that, however, has many good links to useful pages.

StemCells.ca. "The Ethics of Stem Cell Research." Available online. URL: http://www.stemcells.ca. Downloaded on May 27, 2005. A section of the Ethicsweb.ca web site from Canada that deals with various types of ethical issues, stem cell research being one. The site has three main sections: ethics, science, and regulation of stem cell research.

Time.com. "The Stem Cell Debate." Available online. URL: http://www.time.com/time/2001/stemcells. Downloaded on June 3, 2005. A collection of articles and web links about stem cell research and the ethical and legal debate surrounding that issue, current as of late 2001. The web site also includes a major article from the magazine by Richard Lacayo (listed in the Articles section).

Wikipedia. "Stem Cell." Available online. URL: http://en.wikipedia.org/wiki/Stem_cell. Downloaded on May 8, 2005. Wikipedia is a free online encyclopedia with articles on a large number of topics, stem cell research being one. This article provides an excellent general introduction to the scientific background needed to understand SCR issues, along with sections on potential medical applications, ethical issues related to SCR, the international situation, and links to external sites dealing with stem cell issues.

Wired.com. Available online. URL: http://www.wired.com. Downloaded on May 28, 2005. A web site that provides access to current articles on a wide variety of topics, including stem cell research. A recent search, using the site's search function, produces more than 200 articles dealing with the science and technology of stem cell research, legal and regulatory issues, and ethical and moral questions about SCR.

Yahoo! News. "Stem Cell Research." Available online. URL: http://dailynews.yahoo.com/fc/Science/Stem_Cell_Research. Downloaded on May 9, 2005. Yahoo!'s web site that summarizes news articles, editorials and opinions, feature articles, and related web sites on SCR. An excellent source for recent information and opinion on the subject.

SCIENTIFIC AND TECHNICAL TOPICS

BOOKS

A number of technical volumes summarizing the state of the art in the applications of stem cell research in regenerative medicine for specialists in the field are now available. Only a sample of such books are listed below.

2005 Guide to Stem Cell Research Companies. Leesburg, Va.: Data Research, Inc., 2005. A document profiling 82 companies involved in stem cell research as of the date of publication.

Bongso, Ariff, and Eng Hin Lee. *Stem Cells: From Benchtop to Bedside.* Singapore: World Scientific Publishing Company, 2005. A collection of articles with information on the latest research on all types of stem cells and their applications in medicine. The publisher explains that the book is different from similar volumes in that the papers are by "giants in the respective fields."

Chiu, Arlene Y., and Mahendra S. Rao, eds. *Human Embryonic Stem Cells.* Washington, D.C.: AACC Press, 2003. Although this book deals primarily with advances in the study of human embryonic stem cells and their therapeutic applications, it begins with three chapters on legal and ethical issues.

Daley, George Q., Margaret A. Goodell, and Evan Y. Snyder. *Realistic Prospects for Stem Cell Therapeutics.* American Society of Hematology Education Program Book, 2003. An excellent description of the science of stem cell research written at a level of sophistication that should be understandable to most high school students, with a thorough and clear discussion of the kinds of applications that SCR may have in a variety of medical fields. The book is part of the society's Education Program Book Program, available free at http://www.asheducationbook.org/cgi/content/full/2003/1/398.

Greer, Erik V., ed. *Focus on Stem Cell Research.* Hauppauge, N.Y.: Nova Biomedical Books, 2004. A review of progress in stem cell research and its applications to the treatment of nervous system diseases, diabetes, heart disease, autoimmune diseases, Parkinson's disease, end-stage kidney disease, liver failure, cancer, spinal cord injury, multiple sclerosis, and Alzheimer's disease, as well as ethical and legal considerations related to this research.

———. *Trends in Stem Cell Research.* Hauppauge, N.Y.: Nova Biomedical Books, 2005. A continued update on progress in stem cell research. Topics include differentiation of hematopoietic stem cells, stem cells, strategies for the therapeutic use of adult stem cells in regenerative medicine, collection of blood progenitor cells in children, organogenesis from neural crest stem cells, and bone marrow failure in autoimmune diseases.

Habib, Nagy A., Myrtle Y. Gordon, Nataša Levicar, and Long Jiao, eds. *Stem Cell Repair and Regeneration.* Singapore: World Scientific Publishing, 2005. A collection of highly technical papers on the current state of research in stem cell research with special attention to its applications in regenerative medicine.

Keating, A., et al, eds. *Regenerative and Cell Therapy: Clinical Advances.* New York: Springer, 2005. A review for specialists on the state of the art in regenerative cell therapy in the fields of cardiology, hematology, pediatrics, neurology, orthopedics, and infectious diseases.

Kiessling, Ann A., and Scott C. Anderson. *Human Embryonic Stem Cells: An Introduction to the Science and Therapeutic Potential.* Sudbury, Mass.: Jones and Bartlett, 2003. A text designed for readers with some background in cell biology, particularly for medical students, physicians, nurses, veterinarians, and other biomedical scientists. Sidebars providing historical background for the subject of SCR are also provided.

Morser, John, and Shin-Ichi Nishikawa. *The Promises and Challenges of Regenerative Medicine.* New York: Springer, 2005. A collection of papers presented at a conference in Kobe, Japan, on October 20–22, 2004. These are technical reports of interest to specialists in the field. Also, there is a review of the conference in John Morser and Fiona M. McDonald, "The Promises and Challenges of Regenerative Medicine, October 20–22, 2004, Kobe, Japan," *Stem Cells*, vol. 23, no. 5, May 2005, pp. 707–709.

Potten, C. S., ed. *Stem Cells*. San Diego: Academic Press, 1997. Published just before the discovery of the human embryonic stem cell, this anthology focuses primarily on the science of plant and hematopoietic stem cells.

Sell, Stewart, ed. *Stem Cells Handbook*. Totowa, N.J.: Humana Press, 2003. A collection of 43 papers on technical topics in the field of stem cell research.

Turksen, Kursad, ed. *Adult Stem Cells*. Totowa, N.J.: Humana Press, 2004. A collection of papers on the current status of knowledge about adult stem cells, with some attention paid to the ethical issues involved in stem cell research in general. A book primarily for individuals with an extended background in biology.

Who's Who in Stem Cell Research—2005. Leesburg, Va.: Data Research, Inc., 2005. Thumbnail sketches of about 700 individuals involved in stem cell research, with basic biographical information about each person listed.

ARTICLES

Dozens of review articles have been written on the possible applications of stem cell transplantation in the treatment of specific disorders and diseases. Space does not permit the listing of more than a small sample of these articles. For information on the progress of SCR in the treatment of any specific disorder or diseases, refer to the web site for the National Center for Biological Information (http://www.ncbi.nlm.nih.gov).

Bjornson, C. R., et al. "Turning Brain into Blood: A Hematopoietic Fate Adopted by Adult Neural Stem Cells in Vivo," *Science*, vol. 283, no. 5401, January 22, 1999, pp. 534–537. A technical report on (successful) efforts to "re-program" stem cells taken from an adult nervous system to act as blood cells. An important article because it provides fodder for those who argue that adult stem cells have greater plasticity than is sometimes acknowledged for them, thus reducing the necessity to use embryonic stem cells in research.

Blau H. M., T. R. Brazelton, and J. M. Weimman. "The Evolving Concept of a Stem Cell: Entity or Function?" *Cell*, vol. 105, no. 7, June 29, 2001, pp. 829–841. A widely quoted and influential paper that discusses how the term *stem cell* should be defined, arguing that the definition depends more on the way a cell behaves and the functions it performs rather than its morphological (structural) characteristics or its source of origin.

Cibelli, Jose B., et al. "Parthenogenetic Stem Cells in Nonhuman Primates," *Science*, vol. 295, no. 5556, February 1, 2002, p. 819. Description of a research breakthrough in which stem cells are produced from the eggs of a female monkey without being fertilized. The authors argue that the production of stem cells by this mechanism could greatly reduce the ethical debate over funding for stem cell research since the destruction of embryos is not involved.

Committee on Pediatric Research and Committee on Bioethics, American Academy of Pediatrics. "Policy Statement: Human Embryo Research." *Pediatrics*, vol. 108, no. 3, September 2001, pp. 813–816. Also available online. URL: http://aappolicy.aappublications.org/cgi/content/full/pediatrics%3b108/3/813. The committees explain the scientific and medical value of all kinds of stem cell research and recommend that such research be funded by the federal government, at least partly because federal funding "is morally preferable to the currently unregulated private sector approach."

Conrad, Claudius, and Ralf Huss. "Adult Stem Cell Lines in Regenerative Medicine and Reconstructive Surgery," *Journal of Surgical Research*, vol. 124, no. 2, April 2005, pp. 201–208. The authors suggest that "[m]ultipotent adult stem cells seem to be almost comparable to embryonic stem cells with respect to their ability to differentiate into various tissues in vitro and in vivo," and provide a review of research findings that tend to support this view and demonstrate the long-term potential of adult stem cells for regenerative and reconstructive procedures.

Cowan, Chad A., et al. "Derivation of Embryonic Stem-Cell Lines from Human Blastocysts," *New England Journal of Medicine*. vol. 350, no. 13, March 25, 2004, pp. 1,353–1,356. The authors describe the process by which 17 new human embryonic cell lines have been produced, and they offer these cell lines to other researchers for their use. The experiment is important because, since it was conducted with private funds, it provides U.S. researchers with an alternative to the limited number of stem cell lines that can be studied with federal funds, based on President George W. Bush's 2001 administrative policy on SCR.

Daar, Abdallah S., and Lori Sheremeta. "The Science of Stem Cells: Ethical, Legal and Social Issues," *Experimental and Clinical Transplantation*, vol. 1, no. 2, December 2003, pp. 139–146. A general discussion of embryology and embryonic stem cells and a review of recent research developments that bear on the legal, social, and ethical issues that arise with regard to SCR.

Denker, Hans-Werner. "Early Human Development: New Data Raise Important Embryological and Ethical Questions Relevant for Stem Cell Research," *Naturwissenschaften*, vol. 91, no. 1, January 2004, pp. 1–21. A review of scientists' changing understanding of the nature and development of the embryo, providing information that alters the ethical parameters of the current debate over the morality of stem cell research.

Dimmeler, Stefanie, Andreas M. Zeiher, and Michael D. Schneider. "Unchain My Heart: The Scientific Foundations of Cardiac Repair," *Journal of Clinical Investigation*, vol. 115, no. 3, March 2005, pp. 572–583. A review of research in which proteins, cells, and other materials have been used to treat cardiac diseases.

Donovan, Peter J., and John Gearhart. "The End of the Beginning for Pluripotent Stem Cells," *Nature*, vol. 414, no. 6859, November 1, 2001, pp. 92–97. A review of the scientific characteristics of pluripotent stem cells that account for the current excitement about their potential for use in medical therapies.

Fang, Z. F., et al. "Human Embryonic Stem Cell Lines Derived from the Chinese Population," *Cell Research*, vol. 15, no. 5, May 2005, pp. 394–400. Reports by a group of Chinese scientists of the production of six stem cell lines from blastocysts taken from members of the Chinese population.

Fuchs, Elaine, and Julia A. Segre. "Stem Cells: A New Lease on Life," *Cell*, vol. 100, no. 1, January 7, 2000, pp. 143–155. A superb, complete, and clear summary of the characteristics and medical potential of stem cells. One of the best articles from which to obtain a general scientific overview of stem cell research.

Garry, Daniel J., et al. "Ponce De Leon's Fountain: Stem Cells and the Regenerating Heart," *American Journal of the Medical Sciences*, vol. 329, no. 4, April 2005, pp. 190–201. Heart failure continues to be one of the most serious medical problems in developed countries and one that may be treatable by the use of transplanted stem cells. This article reviews some of the medical information relevant to this issue.

Harris, R. G., et al. "Lack of Fusion Requirement for Development of Bone Marrow-derived Epithelia." *Science*, vol. 305, no. 5681, July 2, 2004, pp. 90–93. A technical article describing the failure of getting adult stem cells to differentiate in the body, adding "fuel to the fire" as to whether adult stem cells can be used for medical applications, as hoped by many experts.

Harrower, Timothy, and Roger A. Barker. "Cell Therapies for Neurological Disease—From Bench to Clinic to Bench," *Expert Opinion on Biological Therapy*, vol. 5, no. 3, March 2005, pp. 289–291. A review of some earlier approaches to the treatment of diseases of and damage to the nervous system, and some evidence that stem cells may provide an important breakthrough in this field of therapy.

Haruta, Masatoshi. "Embryonic Stem Cells: Potential Source for Ocular Repair," *Seminars in Ophthalmology*, vol. 20, no. 1, January–March 2005, pp. 17–23. A discussion of the possible use of fetal and embryonic stem cells in the treatment of a variety of ocular disorders and diseases.

Heng, Boon Chin, Hua Liu, and Tong Cao. "Transplanted Human Embryonic Stem Cells as Biological 'Catalysts' for Tissue Repair and Regeneration," *Medical Hypotheses*, vol. 64, no. 6, June 2005, pp. 1,085–1,088. A commentary on recent discoveries that enzymes released by transplanted stem cells may contribute to the regeneration of damaged or injured tissues, a fairly new and promising breakthrough in the potential for stem cell research.

Hochedlinger, Konrad, and Rudolf Jaenisch. "Nuclear Transplantation, Embryonic Stem Cells, and the Potential for Cell Therapy," *New England Journal of Medicine*, vol. 349, no. 3, July 17, 2003, pp. 275–286. A thorough review of the science and technology of stem cell research and its potential applications in medical therapy. The article describes the state of the art of nuclear cloning, therapeutic potential of nuclear transplantation, limitations, and alternatives, and the authors' conclusions.

Hübner, K., et al. "Derivation of Oocytes from Mouse Embryonic Stem Cells." *Science*, vol. 300, no. 5623, May 23, 2003, pp. 1,251–1,256. Researchers at the University of Pennsylvania report that they were able to transform mouse stem cells into mouse egg cells easily. If such a procedure can be reproduced with human stem cells, a major hindrance to the use of human stem cell research might be removed because human embryos would no longer have to be harvested in order to conduct stem cell research.

Hwang, Woo Suk, "Evidence of a Pluripotent Human Embryonic Stem Cell Line Derived from a Cloned Blastocyst," *Science*, vol. 303, no. 5664, March 12, 2004, pp. 1,669–1,674. The first report of the successful cloning of a human embryo by a group of South Korean scientists. Of 176 nuclei transferred to donor egg cells, 30 embryos developed to the blastocyst stage, after which researchers removed 20 embryos to be used for a single new embryonic stem cell line (but see p. 102 of this book).

Irving, Diane. "When Do Human Beings Begin? Scientific Myths and Scientific Facts," *International Journal of Sociology and Social Policy*, vol. 19, nos. 3–4, Fall–Winter 1999, pp. 22–46. An extended and detailed analysis of the process by which fertilization, growth, and development occur in humans and a strong argument that human life begins at the moment of conception. The author offers 14 "myths" that are commonly presented as scientific facts about this process by supporters of embryonic stem cell research, "myths" that she proceeds to show are incorrect.

Kaji, Eugene H., and Jeffrey M. Leiden. "Gene and Stem Cell Therapies," *JAMA*, vol. 285, no. 5, February 7, 2001, pp. 545–550. An excellent general introduction to the science and technology of stem cell research with some attention to the ethical issues surrounding such research.

Kim, J. H., et al. "Dopamine Neurons Derived from Embryonic Stem Cells Function in an Animal Model of Parkinson's Disease," *Nature*, vol. 418, no. 6893, July 4, 2002, pp. 50–56. An important research report because it describes how the successful transplantation of embryonic stem cells into neuronal tissue results in the production of dopamine, a substance whose deficiency in human brains results in the development of Parkinson's disease.

Kurtzberg, Joanne, et al. "Placental Blood as a Source of Hematopoietic Stem Cells for Transplantation into Unrelated Recipients," *New England Journal of Medicine*, vol. 335, no. 3, July 18, 1996, pp. 157–166. A pioneering study on the use of stem cells taken from placental blood in a therapeutic application. The researchers conclude that "placental blood from unrelated donors is an alternative source of stem cells for hematopoietic reconstitution in children."

Leri, Annarosa, Jan Kajstura, and Piero Anversa. "Identity Deception: Not a Crime for a Stem Cell," *Physiology*, vol. 20, no. 3, June 2005, pp. 162–168. A discussion of the plasticity of adult stem cells, which the authors call "the most common and questioned mechanism of growth and repair." They conclude that the most plastic of adult stem cells are bone marrow progenitor cells.

Lewis, Ricki. "A Paradigm Shift in Stem Cell Research?" *The Scientist*, vol. 14, no. 5, March 6, 2000, p. 1. A summary and discussion of recent research that suggests that adult stem cells may be more plastic than had previously been considered and that they may be as useful as embryonic stem cells in medical therapies.

Lin, Haifan. "The Stem Cell Niche Theory: Lessons from Flies," *Nature Reviews Genetics*, vol. 3, no. 12, December 2002, pp. 931–940. A report on research with *Drosophila melanogaster*, a variety of fruit fly, that provides essential information on the structure of the stem cell niche, the microenvironment that directs the replication and differentiation of stem cells, with a discussion of the application of this work to an understanding of the stem cell niche in humans.

Lysaght, Michael J., and Anne L. Hazlehurst. "Private Sector Development of Stem Cell Technology and Therapeutic Cloning," *Tissue Engineering*, vol. 9, no. 3, June 2003, pp. 555–561. Report of a survey taken to discover the role of private funding for stem cell research. The authors found more than 30 firms in 11 countries employing about 1,000 scientists, with 40 percent of this research taking place outside of the United States.

McMahon, Catherine A., et al. "Embryo Donation for Medical Research: Attitudes and Concerns of Potential Donors," *Human Reproduction*, vol. 18, no. 4, April 2003, pp. 871–877. Results of a survey conducted to discover how prospective donors of embryos would feel about the use of such embryos for medical research.

Muraro, Paolo, Riccardo Cassiani-Ingoni, and Roland Martin. "Using Stem Cells in Multiple Sclerosis Therapies," *Cytotherapy*, vol. 6, no. 6, December 2004, pp. 615–620. A review of the results of some trials in which stem cells have been used to treat multiple sclerosis.

Pederson, Roger A. "Embryonic Stem Cells for Medicine," *Scientific American*, vol. 280, no. 4, April 1999, pp. 68–73. A general introduction to the science of stem cell culturing and extraction and a review of some applications that stem cell research may have in the field of medicine.

Ponsaerts, Peter, et al. "Current Challenges in Human Embryonic Stem Cell Research: Directed Differentiation and Transplantation Tolerance," *Journal of Biological Regulators and Homeostatic Agents*, vol. 18, no. 3–4, July–December 2004, pp. 347–351. A technical article dealing with two critical problems in the field of stem cell research, one involving the discovery of factors that cause the differentiation of stem cells and methods for mastering and using those factors, and the other, developing methods for dealing with immune responses by host organisms to transplanted stem cells from foreign bodies.

Rahaman, Mohamad N., and Jeremy J. Mao. "Stem Cell–based Composite Tissue Constructs for Regenerative Medicine," *Biotechnology and Bioengineering*, vol. 91, no. 3, May 31, 2005, pp. 261–284. A review article that discusses the uses of stem cells in the production of synthetic tissues used for implantation in humans and other animals.

"Regenerative Medicine," *Proceedings of the National Academy of Sciences*, vol. 100, supp. 1, September 30, 2003. A special issue of the *Proceedings*, containing a useful two-page overview of the status of stem cell research and 17 colloquium papers reporting on advances in the field.

Rhind, Susan M., et al. "Human Cloning: Can It Be Made Safe?" *Nature Reviews Genetics*. vol. 4, no. 11, November 2003, pp. 855–864. An article that reviews research on efforts to clone adult cells by transferring the nucleus of one cell into a second cell whose nucleus has been removed. The authors say that such experiments for the purpose of producing a new human are unethical, but ask whether the same experiments can be made safe for the purpose of therapeutic cloning.

Shamblott, Michael J., et al., "Derivation of Pluripotent Stem Cells from Cultured Human Primordial Germ Cells," *Proceedings of the National Academy of Sciences*, vol. 95, no. 23, November 10, 1998, pp. 13,726–13,731. One of two classic papers on embryonic stem cell research (along with Thomson, below), in which the extraction of human stem cells from an embryo and demonstration of their pluripotent qualities are demonstrated.

Sonntag, Kai-Christian, Rabi Simantov, and Ole Isacson. "Stem Cells May Reshape the Prospect of Parkinson's Disease Therapy," *Brain Research Molecular Brain Research*, vol. 134, no. 1, March 24, 2005, pp. 34–51. An analysis of the ways in which fetal and stem cell cultures may radically change the treatment of certain neural disorders, such as Parkinson's disease.

Thomson, James A. "Embryonic Stem Cell Lines Derived from Human Blastocysts," *Science*, vol 282, no. 5391, November 6, 1998, pp. 1,145–1,147. One of two classic papers on embryonic stem cell research (along with Shamblott, above) describing the process by which stem cells are obtained and identified from a human embryo.

Verfaillie, Catherine M., Martin F. Pera, and Peter M. Lansdorp. "Stem Cells: Hype and Reality," *Hematology*, 2002, pp. 369–391. Also available online. URL: http://www.asheducationbook.org/cgi/content/full/2002/1/369. A three-part presentation that purports to present current information on stem cells. The three parts discuss (1) the structure and function of telomerase, an enzyme that is involved in the restoration of chromosomes during cell division, (2) current understandings of mammalian pluripotent embryonic stem cells, and (3) current understandings of the plasticity of adult stem cells.

Wagers, Amy J., et al. "Little Evidence for Developmental Plasticity of Adult Hematopoietic Stem Cells," *Science*, vol. 297, no. 5590, September 27, 2002, pp. 2,256–2,259. The authors report on research designed to discover how readily hematopoietic cells differentiate into cells of other types, such as nerve, muscle, or skin cells. They had little success in producing such effects. However, see the response to this article in Neil D. Thiese, et al., "Comment on 'Little Evidence for Developmental Plasticity of Adult Hematopoietic Stem Cells,'" *Science*, vol. 299, no. 5611, February 28, 2003, p. 1,317.

Watt, Fiona M., and Brigid L. M. Hogan. "Out of Eden: Stem Cells and Their Niches," *Science*, vol. 287, no. 5457, February 25, 2000, pp. 1,427–1,430. An interesting discussion of the extrinsic and intrinsic factors that affect the differentiation of a stem cell into a specific type of tissue cell, one of the most difficult problems confronting stem cell researchers.

Watt, Suzanne M., and Marcela Contreras. "Stem Cell Medicine: Umbilical Cord Blood and its Stem Cell Potential," *Seminars in Fetal and Neonatal Medicine*, vol. 10, no. 3, June 2005, pp. 209–220. A comprehensive review of recent research on the use of hematopoietic stem cells from umbilical cord blood for the transplantation of patients with blood disorders.

Wobus, Anna M., and Kenneth R. Boheler. "Embryonic Stem Cells: Prospects for Developmental Biology and Cell Therapy," *Physiological Reviews*, vol. 85, no. 2, April 2005, pp. 635–678. An extensive review of research on the propagation and differentiation of both mouse and human embryonic stem cells, with some consideration of their possible applications in therapeutic medicine.

Zhang, Yan Ding. "Making a Tooth: Growth Factors, Transcription Factors, and Stem Cells," *Cell Research*, vol. 15, no. 5, May 2005, pp. 301–316. The authors review current knowledge of the process by which a tooth grows and explain how developments in stem cell research suggest that it will make possible "the realization of human tooth regeneration in the near future."

Zipori, Dov. "The Stem State: Plasticity Is Essential, Whereas Self-renewal and Hierarchy Are Optional," *Stem Cells*, vol. 23, no. 6, June 2005, pp. 719–726. A review of the traditional definition of stem cells that states they must be not only plastic (that is, capable of differentiating into a variety of different types of specialized cells), but also essentially infinitely self-renewing. The author argues that the former requirement is essential in stem cell applications, but the latter is not.

REPORTS AND STATEMENTS

A number of scientific societies and patient advocacy groups have adopted position statements that describe the progress of stem cell research and potential medical benefits that are likely to accrue from such research. The section below includes only a sample of such statements. For more examples of such statements, consult the web sites of specific organizations or societies.

American Society for Blood and Marrow Transplantation, American Association of Blood Banks, Foundation for the Accreditation of Cellular Therapy, International Bone Marrow Transplant Registry/Autologous Blood and Marrow Transplant Registry, International Society for Cellular Therapy, and National Marrow Donor Program. "ASBMT Position Statement," *Biology of Blood and Marrow Transplantation*, vol. 10, no. 4, April 2004, pp. 283–284. A group of societies working in the field of blood and bone marrow transplantation outline the justifications for the federal support of stem cell research.

Human Stem Cell Research. Canberra: Australian Academy of Science, April 18, 2001. Available online. URL: http://www.science.org.au/reports/stemcell.pdf. Downloaded on June 5, 2005. The academy considers progress in the field of stem cell research, potential medical applications, Australia's current attitudes toward stem cell research, and possible legislative action that needs to be taken with regard to SCR.

Juvenile Diabetes Research Foundation. "Why Federal Stem Cell Policy Must Be Expanded: A JDRF Scientific White Paper: August 2004." Available online. URL: http://www.jdrf.org/files/About_JDRF/JDRF%20Stem%20Cell%20White%20Paper%202004.pdf. Posted in August 2004. An extended analysis of existing federal policy on stem cell research, with reasons for the inadequacies of this policy and recommendations for extended the federal funding options to be permitted.

Select Committee on Stem Cell Research. *Stem Cell Research.* London: United Kingdom Parliament, House of Lords, February 13, 2002. A very complete overview of the science of stem cell research and its potential in medical therapeutics and other fields of science, with a section on the legal status of SCR in the United Kingdom.

Stem Cells: Scientific Progress and Future Research Directions. Washington, D.C.: Department of Health and Human Services, June 2001. Available online. URL: http://stemcells.nih.gov/info/scireport. A report prepared in response to a request by Secretary of Health and Human Services Tommy Thompson on the state of scientific research on stem cells. The report covers a variety of technical topics, including the characteristics of adult and embryonic stem cells; the use of stem cells to treat a variety of diseases, such as autoimmune diseases, diabetes, disorders of the nervous system, and cardiac conditions; the safety of stem cell research; and the use of genetically modified stem cells in experimental gene therapies.

Varmus, Harold. "Statement on Stem Cell Research." Before the Senate Appropriations Subcommittee on Labor, Health and Human Services, Education and Related Agencies, January 26, 1999. Also available online. URL: http://www.hhs.gov/asl/testify/t990126a.html. The director of the National Institutes of Health reports to the Congress on the scientific background and possible medical applications of stem cell research.

WEB DOCUMENTS

American Society of Hematology. "ASH Supports Both Adult and Embryonic Stem Cell Research." Available online. URL: http://www.hematology.org/government/policy/stemcell_042602.cfm. Downloaded on June 5, 2005. The society explains why it supports federal funding of embryonic and adult stem cell research and somatic cell nuclear transfer (SCNT).

BioethicsWeb. Available online. URL: http://bioethicsweb.ac.uk. An outstanding web site that contains a vast array of Internet links to sites on a wide variety of bioethics topics, primarily in the fields of cloning, genetics, and stem cell research. Researchers can search the web site by way of a number of categories, including biomedical subject matter areas; clinical practice; countries and regions; environment, agriculture, and foods; ethics: theory and concepts; organizations; people; reference materials; research conduct; and society, policy, and law.

Biotechnology Industry Organization. "BIO Response to NBAC on Stem Cell Research." Available online. URL: http://www.bio.org/bioethics/background/NBAC.asp. Posted on March 22, 1999. A letter from the Biotechnology Industry Organization to the chair of the National

Bioethics Advisory Commission, in response to NBAC's report on stem cell research. The letter outlines BIO's stance on stem cell research and its potential applications in the field of therapeutic medicine and other scientific areas.

Boston University. "Bioethics Working Group @ Boston University." Available online. URL: http://www.bu.edu/bioethics/pages/stemcells.html. Downloaded on June 13, 2005. A good general introduction to the science of stem cell research and the political, legal, and ethical issues related to SCR. The site consists of links to essential primary documents in the categories of "Overview," "Web Resources," and "Statements by George W. Bush."

Cord-Blood.org. Available online. URL: http://www.cord-blood.org. Downloaded on May 21, 2005. A web site that discusses the use of umbilical cord blood as a source of stem cells, with a discussion of the science involved and the pros and cons of "banking" one's cord blood during pregnancy.

Data Trends, Inc. "Stem Cell Research News." Available online. URL: http://www.stemcellresearchnews.com. Downloaded on June 2, 2005. A web site founded in October 1999 for the purpose of providing in-depth information on scientific advances and legislative and regulatory activities affecting SCR. Access to some articles is by subscription only, although other information and reports are available free to online users.

Dolan DNA Learning Center. "How Embryonic Stem Cell Lines Are Made." Available online. URL: http://www.dnalc.org/stemcells.html. Downloaded on May 14, 2005. An excellent animated explanation of the process by which stem cells are harvested from an embryo.

Genetic Science Learning Center, University of Utah. "Stem Cells in the Spotlight." Available online. URL: http://gslc.genetics.utah.edu/units/stemcells. Downloaded on May 17, 2005. An instructional web site created for younger students with excellent animated explanations of topics such as what is a stem cell, what are some different types of stem cells, what is the goal of stem cell research, and stem cell therapies today.

Genome News Network. "Stem Cells." Available online. URL: http://www.genomenewsnetwork.org/categories/index/stemcells.php. Downloaded on May 22, 2005. Genome News Network is published by the J. Craig Venter Institute, in Rockville, Maryland, a not-for-profit institute founded in 2002, dedicated to providing information on genome-related issues, one of which is stem cell research. The web site contains many up-to-date articles on the current status of research on stem cells.

The Harvard Stem Cell Institute. "Unlocking the Promise of Stem Cells." Available online. URL: http://athome.harvard.edu/programs/psc/index.html. Downloaded on June 2, 2005. The report of an interactive videoconference hosted by the Harvard Club of New York City on the occasion of the establishment of the Harvard Stem Cell Institute.

The Lasker Foundation. "Stem Cell Research." Available online. URL: http://www.laskerfoundation.org/rprimers/stemcell/stemcell.html. Downloaded on May 5, 2005. A page on the Lasker Foundation web site that provides a general introduction to the science of stem cells, a profile of scientist John D. Gearhardt, some opinions and quotes about stem cell research, and sources of additional information.

Medline Plus. "Stem Cells and Stem Cell Transplantation." Available online. URL: http://www.nlm.nih.gov/medlineplus/stemcellsandstemcell transplantation.html. Downloaded on May 10, 2005. A service of the National Library of Medicine and the National Institutes of Health, this web site is a large database that includes news, health resources, clinical trials, and additional information on stem cell research.

National Center for Biological Information. Available online. URL: http://www.ncbi.nlm.nih.gov. Downloaded on June 5, 2005. A web site sponsored by the National Institutes of Health and the National Library of Medicine that indexes countless numbers of articles about medical and health-related topics, including stem cell research. The site is probably the first place a researcher should go to obtain a complete listing of articles on the scientific, technological, political, legal, ethical, and social aspects of any such topics. It almost certainly provides the most complete guide to current developments in scientific research on stem cells.

National Institutes of Health. "Information on Eligibility Criteria for Federal Funding of Research on Human Embryonic Stem Cells." Available online. URL: http://stemcells.nih.gov/research/registry/eligibilityCriteria.asp. Downloaded on May 12, 2005. A listing of the organizations, agencies, and institutions that own the 78 embryonic stem cell lines that meet the criteria established by President George W. Bush on August 9, 2001, for federal funding of SCR.

National Library of Medicine. "PubMed." Available online. URL: http://www.ncbi.nlm.nih.gov/entrez/query.fcgi?db=PubMed&itool=toolbar. Downloaded on May 1, 2005. A very large database that includes articles from thousands of scientific journals. One can search for terms such as "stem cells" to find all articles that include that term anywhere in the body or abstract of the article. As of late 2005, such a search produced more than 60,000 articles.

National Public Radio. "Stem Cell Therapy for Heart Patients." Available online. URL: http://www.npr.org/templates/story/story.php?storyId=1863904. Posted on April 30, 2004. A discussion conducted on NPR's Talk of the Nation program on April 30, 2004, featuring heart surgeon Amit Patel of the University of Pittsburgh's School of Medicine.

ScienceDaily. Available online. URL: http://www.sciencedaily.com. Downloaded on May 1, 2005. A web site that contains a very complete coverage of news on virtually every scientific topic, including stem cell research. Use the site's search function to find a list of recent news articles on developments in the field of SCR. Articles on related topics, such as the legal and political implications of SCR, are also available in other sections of the web site, such as government or internal policy.

Smith, Chris. "The CytoMatrix: Reloaded." Naked Scientists. Available online. URL: http://nakedscientists.com/HTML/Columnists/chrissmithcolumn2.htm. Downloaded on May 16, 2005. Describes a new technique for growing stem cells outside the body.

"Stem Cell Web Focus." Available online. URL: http://www.nature.com/nature/stemcells/index.html. Downloaded on June 20, 2005. An Internet reflection of the June 20, 2002, issue of the journal *Nature* that offers a number of important scientific papers on stem cell research as well as links to articles from earlier issues and a selection of classic papers on stem cells.

"Therapeutic Uses of Stem Cells for Spinal Cord Injuries: A New Hope." Available online. URL: http://www.namiscc.org/newsletters/December01/SCI-stem-cell-research.htm. Downloded on June 15, 2005. A page on the web site of the National Alliance for the Mentally Ill of Santa Cruz County that provides an extensive update on progress that has been made in the use of stem cells for treating spinal cord injuries.

Verter, Francis. "A Parent's Guide to Choosing a Private Cord Blood Bank." Available online. URL: http://www.parentsguidecordblood.com. Downloaded on June 3, 2005. A web site that explains the potential use of umbilical cord blood as a source of stem cells, along with a discussion of the medical pros and cons of banking of cord blood; diseases that can or may be treated by blood stem cells; types of cord blood banks; ongoing research; news reports; and ethical, legal, and privacy issues.

WebMedHealth. "Stem Cells: What They Are and Why They're Controversial." Available online. URL: http://my.webmd.com/content/pages/5/1728_86999?z=1626_00000_5022_pe_02. Downloaded on May 23, 2005. A clear, well-written and comprehensive description of what stem cells are, what medical potential they may hold, and what the issues are surrounding their use.

The White House. "Stem Cell Fact Sheet." Available online. URL: http://www.whitehouse.gov/news/releases/2001/08/20010809-1.html. Downloaded on May 20, 2005. A web site that outlines the types of stem cell research that can be funded in the United States, based on President George W. Bush's policy statement of August 9, 2001.

HISTORY

BOOKS

Maienschein, Jane. *Whose View of Life?: Embryos, Cloning, and Stem Cells.* Cambridge, Mass.: Harvard University Press, 2003. The author provides a very useful general introduction to the topic of stem cell research, but focuses on the variety of opinions that have been expressed throughout the centuries as to the meaning of human life and how current beliefs affect the present state of stem cell research in the United States and elsewhere.

Parson, Ann B. *The Proteus Effect: Stem Cells and Their Promise for Medicine.* Washington, D.C.: Joseph Henry Press, 2004. A well-written work on the history of stem cell research with a good overview of the topic and the contributions it may make to regenerative medicine and other fields of research.

ARTICLES

Andrews, Peter W. "From Teratocarcinomas to Embryonic Stem Cells," *Philosophical Transactions of the Royal Society of London*, vol. 357, no. 1420, April 29, 2002, pp. 405–417. A detailed, if somewhat technical, review of the development of stem cell research, with special attention to its application to regenerative medicine.

Bongso, Ariff, and Mark Richards. "History and Perspective of Stem Cell Research," *Best Practice & Research in Clinical Obstetrics & Gynaecology*, vol. 18, no. 6, December 2004, pp. 827–842. Background of discoveries of embryonic, germ, adult, and fetal stem cells, the scientific status of current research, and some issues related to progress in SCR.

Boonstra, Heather. "Human Embryo and Fetal Research: Medical Support and Political Controversy," *The Guttmacher Report on Public Policy*, vol. 4, no. 1, February 2001, pp. 4–7. Also available online. URL: http://www.guttmacher.org/pubs/tgr/04/1/gr040103.html. A good review of government policy in the United States on stem cell research and related issues of human experimentation.

Center for Practical Bioethics. "Early Stem Cell Research," issue 1, Spring 2005. Also available online. URL: http://www.practicalbioethics.org/050405/FINALStemCellBrief032205.pdf. A newsletter that presents a general overview of stem cell research, with a brief, but useful history of the topic on page 4.

Childress James F. "Human Stem Cell Research: Some Controversies in Bioethics and Public Policy," *Blood Cells, Molecules, and Diseases*, vol. 32, no. 1, January–February 2004, pp. 100–105. A review of the ethical and policy issues over stem cell research as they have arisen within the work of two national committees, the National Bioethics Advisory Commission and the President's Council on Bioethics.

Cooper, Melinda. "Regenerative Medicine: Stem Cells and the Science of Monstrosity," *Journal of Medical Humanities*, vol. 30, no. 1, June 2004, pp. 12–22. The subject of medical "monstrosities" was a topic of considerable interest to 19th-century researchers. The author explores the philosophical and ethical contexts within which that field of research has some relationship to the present-day dispute over stem cell research in terms of the effort to provide a specific and useful meaning for "human life."

Edwards, Robert G. "IVF and the History of Stem Cells," *Nature*, vol. 413, no. 6854, September 27, 2001, pp. 349–351. The author, a pioneer in in vitro fertilization technology, argues that embryo stem cells have been used in research for decades. He argues that the ethical issues of SCR are considerably less important than those surrounding other medical procedures, such as cloning, preimplantation diagnosis, early human embryo research, and gamete donation.

Golde, David W. "The Stem Cell," *Scientific American*, vol. 265, no. 12, December 1991, pp. 36–43. A description of hematopoietic stem cells and their function in the human body and possible applications in therapeutic medicine, written at a time before the discovery of human embryonic stem cells and the rise of the debate over their possible medical applications.

Green, Ronald M. "Determining Moral Status," *American Journal of Bioethics*, vol. 2, no. 1, Winter 2002, pp. 20–30. Some thoughts on the ethical issues that arose in the early years of the stem cell research debate with suggestions as to how some general moral principles may be applied.

Hopkins, Patrick D. "Bad Copies. How Popular Media Represent Cloning as an Ethical Problem," *Hastings Center Report*, vol. 28, no. 2, March–April 1998, pp. 6–13. An article that looks at the issues raised by cloning successes in the mid- to late-1990s, with relevance to issues related to stem cell research. The author argues that the media, to a large extent, define the boundaries surrounding many socioscientific issues, and, in the case of cloning experiments, have focused on three major concerns: "the loss of human uniqueness and individuality, the pathological motivations of a cloner, and the fear of out-of-control scientists."

Keller, G. "Embryonic Stem Cell Differentiation: Emergence of a New Era in Biology and Medicine," *Genes and Development*. vol. 19, no. 10, May 15, 2005, pp. 1,129–1,155. A review of the history of stem cell research dating to the 1980s, with commentary on the way SCR has contributed to scientists' understanding of cells and development and the potential contribution it may make to medical progress.

Laslett, Andrew L., Adam A. Filipczyk, and Martin F. Pera. "Characterization and Culture of Human Embryonic Stem Cells," *Trends in Cardiovascular Medicine*, vol. 13, no. 7, October 2003, pp. 295–301. A review of research papers on the subject of stem cells published since the 1960s.

Lensch, M. William, and George Q. Daley. "Origins of Mammalian Hematopoiesis: In Vivo Paradigms and In Vitro Models, *Current Topics in Developmental Biology*, vol. 60, 2004, pp. 127–196. A somewhat technical articles that includes some major steps taken over the past half century in the study of embryonic stem cells.

Lewis, Ricki. "A Stem Cell Legacy: Leroy Stevens," *The Scientist*, vol. 14, no. 5, pp. 19–24. A review of the life and work of one of the major pioneers of stem cell research.

Maienschein, Jane. "What's in a Name: Embryos, Clones, and Stem Cells," *American Journal of Bioethics*, vol. 2, no. 1, Winter 2002, pp. 12–19. Among the problems that developed during the early stages of the debate over stem cell research was terminology. The author asks whether politicians, reporters, legislators, and the general public really understood the terms that were being used in discussing SCR issues and what that possible lack of understanding of terminology might mean for the debate itself.

Maroney, Helen M. "Bioethical Catch-22: The Moratorium on Federal Funding of Fetal Tissue Transplantation Research and the NIH Revitalization Amendments," *Journal of Contemporary Health Law Policy*, vol. 9, Spring 1993, pp. 485–519. An extended discussion of the moratorium on fetal tissue transplantation research (FTTR) in 1988, the passage of the NIH Revitalization Amendments in 1993, the veto by President George H. W. Bush of this act, and the consequences of this political maneuvering for FTTR research in the United States.

Messikomer, Carla M., Renée C. Fox, and Judith P. Swazey. "The Presence and Influence of Religion in American Bioethics," *Perspectives in Biology and Medicine*, vol. 44, no. 4, Fall 2001, pp. 485–508. Acknowledging the crucial role that religious thought has had in discussions of bioethics since the 1980s, the authors use two case studies, the debate over the meaning of death and the recent work of the National Bioethics Advisory Commission, have been influenced by religious thinkers.

Nierras, Concepcion R., et al. "Human Embryonic Stem Cell Research and the Juvenile Diabetes Research Foundation International—A Work in Progress," *Pediatric Diabetes*, vol. 5, suppl. 2, December 2004, pp. 94–98. A description of the work of the Diabetes Research Foundation and its support for stem cell research as a potential cure for the disease.

"Stem Cell: Hope or Hype?" *BiC News*, issue 8, January 1, 2005. Also available online. URL: http://www.bic.org.my/BICnews/BICnews8.pdf. One of the most complete and informative reviews of the chronology of stem cell research is available on pages 3 to 6 of this newsletter. Other topics included in the publication are the unique properties of stem cells, classification of stem cells based on origins, classification of stem cells based on their potential, laboratory procedures, potential application of stem cells, recent headlines, obstacles in stem cell research, and ethical issues.

Wertz, D. C. "Embryo and Stem Cell Research in the USA: A Political History," *Trends in Molecular Medicine*, vol. 8, no. 3 March 2002, pp. 143–146. A review of the history of legislation and policy dealing with stem cell research dating back to the legalization of elective abortions in 1973, with attention to the unique political and religious factors at work in the United States contributing to this history.

Zoloth, Laurie. "Jordan's Banks: A View from the First Years of Human Embryonic Stem Cell Research," *American Journal of Bioethics*, vol. 2, no. 1, Winter 2002, pp. 3–11. The author discusses ethical issues that arose somewhat unexpectedly in the late 1990s with discoveries in the field of human embryonic stem cell research, with suggestions that many points still need to be explored with respect to ethical issues involved in such research.

Web Documents

"History of Stem Cell Research." All about Popular Issues. Available online. URL: http://www.allaboutpopularissues.org/history-of-stem-cell-research-faq.htm. Downloaded on June 9, 2005. A concise review of the relatively short history of research involving stem cells from humans and other animals.

National Institutes of Health. "Stem Cell Information." Available online. URL: http://stemcells.nih.gov/index.asp. Downloaded on May 12, 2005. A comprehensive survey of information about stem cell research and the issues surrounding those studies, with pages covering basic information, ethical issues, federal policy, research topics, recent news and announcements, and a very useful glossary.

National Marrow Donor Program. "History of Stem Cell Transplants." Available online. URL: http://www.marrow.org/NMDP/history_stem_cell_transplants.html. Downloaded on June 8, 2005. A historical review of the transplantation of bone marrow cells for the treatment of disease, primarily leukemias.

"A Short History of Pluripotent Cell Investigation." Pluripotent Cell Biology. Available online. URL: http://daley.med.harvard.edu/History.htm. Downloaded on June 10, 2005. A somewhat technical review of some major events in the study of embryonic stem cells.

MSNBC.com. "Sick Dog Gets $45,000 Stem Cell Transplant." Available online. URL: http://www.atsnn.com/story/133252.html. Posted on May

26, 2005. A Washington state couple paid $45,000 to have a stem cell implant for their golden retriever with lymphoma. In a response to this story, one correspondent offers a "quick and dirty look at the *public* history of stem cell research" and lists scientific papers relating to the use of stem cell materials for transplantations as far back as the 1950s.

LAWS AND REGULATIONS

BOOKS

Committee on Guidelines for Human Embryonic Stem Cell Research, Board on Life Sciences, National Research Council and Health Sciences Policy Board, Institute of Medicine. *Guidelines for Human Embryonic Stem Cell Research.* Washington, D.C.: National Academies Press, 2005. Although federal policy currently limits the types of embryonic stem cell research that can be conducted, that research continues to expand and develop rapidly, often because it has been funded by private money. This report is intended to provide guidelines for the conduct of such research, whether it be funded by private or public monies. The book includes chapters on "Scientific Background of Human Embryonic Stem Cell Research," "Addressing Ethical and Scientific Concerns Through Oversight," "Current Regulation of Human Embryonic Stem Cell Research," "Recruiting Donors and Banking hES Cells," and "National Academies Guidelines for Research on Human Embryonic Stem Cells." The book also contains an excellent and complete list of references.

ARTICLES

Andrews, Lori B., "State Regulation of Embryo Stem Cell Research," in *Ethical Issues in Human Stem Cell Research*, vol. 2: Commissioned Papers, Washington, D.C.: National Bioethics Advisory Commission, January 2000, pp. A1–A13. The author examines a variety of legal issues surrounding the use of embryonic stem cells in research and reviews the legal status of such research in all 50 states.

Caulfield, Timothy, Lori Knowles, and Eric Meslin. "Law and Policy in the Era of Reproductive Genetics," *Journal of Medical Ethics*, vol. 30, no. 4, August 2004, pp. 414–417. The authors suggest that dealing with complex and contentious issues such as stem cell research by passing laws is not a productive or effective way of handling such issues and that regulatory policies are likely to be more effective. They suggest one possible approach to the debate over SCR based on this philosophy.

Annotated Bibliography

De Trizio, Ella, and Christopher S. Brennan, "The Business of Human Embryonic Stem Cell Research and an International Analysis of Relevant Laws," *Journal of Biolaw & Business*, vol. 7, no. 4, 2004, pp. 14–22. A survey of laws regulating human embryonic stem cell research in various scientifically advanced countries located throughout the Pacific Rim, Europe, and North America. The article attempts to explain the impact these laws have had on governmental and private funding of human embryonic stem cell research.

Dolgin, Janet L. "Embryonic Discourse: Abortion, Stem Cells, and Cloning," *Florida State University Law Review*, vol. 31, no. 1, Fall 2003, pp. 101–162. Also available online. URL: http://www.law.fsu.edu/journals/lawreview/downloads/311/Dolgin.pdf. The author compares the debate over abortion and the utilization of embryonic stem cells in research and finds certain common themes, but also some distinct differences in the way these debates have played themselves out. The author is professor of constitutional law at Hofstra University.

Doring, Ole. "Chinese Researchers Promote Biomedical Regulations: What Are the Motives of the Biopolitical Dawn in China and Where Are They Heading?" *Kennedy Institute of Ethics Journal*, vol. 14, no. 1, March 2004, pp. 39–46. A review of research on stem cells being carried out in China with an analysis of the future significance of this research. The article is followed by a statement by the Ethics Committee of the Chinese National Human Genome Center at Shanghai on "Ethical guidelines for human embryonic stem cell research" (pp. 47–54).

Erwin, Consuelo G. "Embryonic Stem Cell Research: One Small Step for Science or One Giant Leap Back for Mankind?" *University of Illinois Law Review*, no. 1, 2003, pp. 211–243. The author compares current research using stem cells with experimentation on humans by Nazi scientists during World War II and finds that such research "violates the ethical standards and purpose of the Nuremberg Code and should be banned by federal legislation."

Fox, Marie. "Pre-persons, Commodities or Cyborgs: The Legal Construction and Representation of the Embryo." *Health Care Analysis*, vol. 8, no. 2, June 2000, pp. 171–188. A review of the way embryos have been treated historically under British law, with suggestions as to how embryos used in SCR should be considered within this historical context.

Hadaway, Ben. "Embryonic Stem Cell Research Finally Regulated," *CMAJ*, vol. 170, no. 7, March 30, 2004, p. 1086. After a long delay, during which there was considerable debate and discussion, the Canadian government adopts guidelines to regulate stem cell research conducted within the country.

Hansen, Bert. "Embryonic Stem Cell Research: Terminological Ambiguity May Lead to Legal Obscurity," *Medicine and Law*, vol. 23, no. 1, 2004, pp. 19–28. Because governmental agencies are not always clear and consistent in their use of scientific terminology, some of the laws and regulations they draft may be ambiguous and difficult to understand and enforce. The current legal situation in Europe is analyzed as an example of this situation.

Hill, Helen. "Australia's Bio-regulatory Framework: Leading the Way for Stem Cell Research," *Journal of BioLaw and Business*, vol. 7, no. 2, 2004, pp. 61–64. A review of the rapid progress made by biotechnology in general, and stem cell research in particular, in Australia over the past five years.

Jain, K. K. "Ethical and Regulatory Aspects of Embryonic Stem Cell Research," *Expert Opinion on Biological Therapy*, vol. 2, no. 8, December 2002, pp. 819–826. A review of the regulatory status of stem cell research in countries around the world that finds that regulation is most liberal in China, India, Israel, Singapore, Sweden, and the United Kingdom. The author suggests that restrictive policies in the United States may cause this nation to lose its lead in stem cell research to countries with more liberal policies.

Knoepffler, Nikolaus. "Stem Cell Research: An Ethical Evaluation of Policy Options," *Kennedy Institute of Ethics Journal*, vol. 14, no. 1, March 2004, pp. 55–74. The author defines eight distinct "policy options" that a country can develop and then attempts to evaluate each of the eight options. Also see a response to this paper in the same issue of the journal by Alfonso Gomez-Lobo, "On the Ethical Evaluation of Stem Cell Research: Remarks on a Paper by N. Knoepffler" (pp. 75–80).

Kraco, K. "University Seeks Private Funding for Embryonic Stem Cell Research," *Minnesota Medicine*, vol. 87, no. 6, June 2004, p. 11. Also available online. URL: http://www.mmaonline.net/publications/MNMed2004/June/Pulse.html. The University of Minnesota decides to go ahead with stem cell research provided it can find funds from private sources to support such research. State legislators also announce plans to place before voters a proposal for state funding of SCR.

Martin, Patricia A., and Martin L. Lagod. "The Human Preembryo, the Progenitors, and the State: Toward a Dynamic Theory of Status, Rights, and Research Policy," *High Technology Law Journal*, vol. 5, no. 2, Fall 1990, pp. 257–310. Also available online. URL: http://www.law.berkeley.edu/journals/btlj/articles/vol5/Martin/html/text.html. An extended discussion on the legal rights of the human pre-embryo, donors of the preembryo and other embryo- and fetal-like materials, the legal status of both at the time the article was written, and some recommendations for a legal policy that takes into consideration traditional legal and ethical values and the development of a new biological technology.

Miehl, Kathryn L. "Pre Embryos: The Tiniest Speck of Potential Life Carrying the Seeds for Sweeping Change," *Journal of Technology Law and Policy*, vol. 6, no. 1, Fall 2003. Also available online. URL: tlp.law.pitt.edu/articles/Vol6Miehl.pdf. An excellent review of the current status of the embryo as evidenced by state, federal, and international law and relevant cases from decisions rendered by state courts and the U.S. Supreme Court.

Pattinson, Shaun D., and Timothy Caulfield. "Variations and Voids: The Regulation of Human Cloning Around the World," *BMC Medical Ethics*, vol. 5, December 13, 2004, pp. 9–17. Report of a survey on laws and regulations on human cloning in 30 countries around the world. The authors found that there is still much ambiguity about the legal status of stem cell research and that no two countries have the same laws and regulations.

Pennings, Guido. "New Belgian Law on Research on Human Embryos: Trust in Progress Through Medical Science," *Journal of Assisted Reproduction and Genetics*, vol. 20, no. 8, August 2003, pp. 343–346. The author describes and discusses a new law that permits all types of research on embryos intended for therapeutic purposes and for the increase of medical knowledge, including germline and somatic gene therapy, therapeutic cloning, and the development of embryonic stem cell lines. The law does prohibit other types of research conducted for nonmedical reasons, such as reproductive cloning and eugenic practices.

Resnick, David B. "The Commercialization of Human Stem Cells: Ethical and Policy Issues," *Health Care Analysis*, vol. 10, no. 2, June 2002, pp. 127–154. The author suggests that many of the initial scientific and ethical questions about stem cell research are now being answered and that a "second round" of questions involving the commercialization of stem cell research needs to be addressed. He argues that (1) it should be legal to buy and sell embryonic stem cells and stem cell products and to patent stem cells and stem cell products; (2) it should not be legal to buy, sell, or patent human embryos; (3) patents on stem cells and related products should not be overly broad and should be granted only when inventors demonstrate plausible uses for their products; (4) patent exemptions should be granted institutions of higher learning when stem cells and related products are to be used for research; (5) restrictions on patents may be necessary to prevent companies from using their patents to stifle competition; and (6) the issue of property rights should continually be revisited in order to develop regulations that "maximize the social, medical, economic, and scientific benefits of ES cell research and product development."

Revel, Michel. "Human Reproductive Cloning, Embryo Stem Cells and Germline Gene Intervention: An Israeli Perspective," *Medicine and Law*, vol. 22, no. 4, 2003, pp. 701–732. A review of the legal status of embryonic stem cell research in Israel and future considerations in light of the view of Jewish law toward research aimed at curing disease and saving human lives.

Scherer, Ron. "States Race to Lead Stem-cell Research," *Christian Science Monitor*, February 25, 2004, p. 1ff. While the federal government holds off on funding stem cell research, New Jersey, California, and a number of other states are pushing forward to get citizen approval to use state tax dollars to support such research.

Stevens, Denise. "Embryonic Stem Cell Research: Will President Bush's Limitation on Federal Funding Put the United States at a Disadvantage? A Comparison Between U.S. and International Law," *Houston Journal of International Law*, vol. 25, no. 3, Spring 2003, pp. 623–653. Also available online. URL: http://www.hjil.org/Articles/ArticleFiles/25_3_131.pdf. The author surveys the current legal status of stem cell research in the United States and other nations and concludes that the U.S. government should change its existing policies and agree to fund embryonic stem cell research in order to prevent having the United States fall behind other nations scientifically and economically.

Tanne, Janice Hopkins. "US Universities Get Round Regulations on Stem Cell Research," *BMJ*, vol. 328, no. 7448, May 8, 2004, p. 1,094. Available online: URL http://bmj.bmjjournals.com/cgi/content/full/328/7448/1094. A review of some ways that universities are pursuing to avoid the limitations placed on federal funding by President George W. Bush's announcement of August 2001.

Troeger, Melinda. "Comment: The Legal Status of Frozen Pre-embryos When a Dispute Arises During Divorce," *American Academy of Matrimonial Lawyers Journal*, vol. 18, no. 2, July 2004, pp. 563–587. An extensive review of existing federal and state laws and case law dealing with the legal status of pre-embryos.

Varmus, Harold E. "The Challenge of Making Laws on the Shifting Terrain of Science," *Journal of Law and Medical Ethics*, vol. 28, no. 4 (supplement), Winter 2000, pp. S46–S53. A discussion of one of the classic problems in the philosophy of science, to wit, to what extent and how society can and should regulate scientific research. That problem has become more difficult to answer in recent decades as the pace of scientific discoveries increase and the law, which tends to move slowly, is unable to keep up with the pace of scientific advances.

Walters, LeRoy. "Human Embryonic Stem Cell Research: An Intercultural Perspective," *Kennedy Institute of Ethics Journal*, vol. 14, no. 1, March 2004, pp. 3–38. An analysis of policies on stem cell research in four parts of the world and of views expressed at a recent United Nations debate on

SCR. The author finds that religious views have a strong influence on policies adopted by various countries.

Weed, Matthew. "Ethics, Regulation, and Biomedical Research," *Kennedy Institute of Ethics Journal*, December 2004, pp. 361–368. A review of commissions and other groups that have been appointed in the past to provide advice on and/or regulate biomedical research, with a recommendation for the establishment of an independent, autonomous, permanent Federal Life Sciences Policy Commission to deal with such issues in the future. The constitution and operation of the commission should reflect lessons learned from the operation of earlier similar groups.

REPORTS AND STATEMENTS

Duffy, Diane T. "Background and Legal Issues Related to Stem Cell Research." Washington, D.C.: Congressional Research Service, June 12, 2002. Also available online. URL: http://fpc.state.gov/documents/organization/11274.pdf. A review of the legal and regulatory status of stem cell research from the late 1990s to the date of the report.

European Commission. "Survey on Opinions from National Ethics Committees or Similar Bodies, Public Debate and National Legislation in Relation to Human Embryonic Stem Cell Research and Use." Directorate E: Biotechnology, Agriculture and Food, September 2003. A summary of a study done to determine attitudes and regulations about stem cell research in 15 nation members of the European Union. The report consists of verbatim answers to a questionnaire developed by researchers along with a useful table summarizing laws and regulations in each of the countries surveyed.

"Human Pluripotent Stem Cell Research: Guidelines for CIHR-Funded Research." Ottawa, Ontario, Canada: Canadian Institutes of Health Research, March 2002. Available online. URL: http://www.cihr-irsc.gc.ca/e/1487.html. A report providing guidelines for the funding of stem cell research in Canada. The guidelines are especially of interest to U.S. citizens because they provide a basis for comparison and contrast with similar guidelines issued by the U.S. National Institutes of Health in 2001.

Johnson, Judith A. "Human Cloning." Washington, D.C.: Congressional Research Service, December 19, 2001. A review of administrative orders, legislation, regulatory actions, and other documents dealing with the process of human cloning and related procedures in the United States.

"National Institutes of Health (NIH) Update on Existing Human Embryonic Stem Cells." Bethesda, Md.: National Institutes of Health, August 27, 2001. Available online: URL http://stemcells.nih.gov/policy/statements/082701list.asp. A statement of the U.S. government's official position on stem cell research, based on the policies outlined by President George W. Bush in his speech of August 9, 2001.

WEB DOCUMENTS

Agnew, Bruce. "The Politics of Stem Cells." Genome News Network. Available online. URL: http://www.genomenewsnetwork.org/articles/02_03/stem.shtml. Downloaded on May 12, 2005. An overview of the legal status of embryonic stem cell research in the United States and the problems faced by SCR researchers because of current administrative regulations in this country.

Carmack, Tom. "Stemming Designs for Inhibitory Government Regulation: Allowing Stem Cell Research to Achieve its Pluripotential," *Business Law Journal.* Available online. URL: http://blj.ucdavis.edu/article/516. Posted December 1, 2003. The author reviews the scientific and legal status of stem cell research and concludes that the U.S. government should fund such research, at least partly because the partial support currently being provided is inconsistent with the full support available from private sources.

Castle & DeGette Press Releases. Available online. URL: http://www.house.gov/castle/Castle%20DeGette%20Press%20Releases.html. Downloaded on May 2, 2005. About two dozen press releases from the offices of Representatives Mike Castle (R-Del.) and Diana DeGette (D-Colo.), primary sponsors of H.R. 810, a bill designed to expand the use of stem cells in federally funded programs. The bill passed the House of Representatives but was never voted on by the Senate. The press releases are important because they outline arguments for changing the federal government's existing policy on stem cell research.

Eisenberg, Daniel. "Stem Cell Research in Jewish Law." Available online. URL: http://www.jlaw.com/Articles/stemcellres.html. Downloaded on May 27, 2005. An article that attempts to determine how biblical and rabbinical law view stem cell research. In general, the author argues that such research is permissible because it has the potential for saving human lives.

Genome News Network. "Stem Cells: Policies and Players." Available online. URL: http://www.genomenewsnetwork.org/resources/policiesandplayers. Downloaded on May 20, 2005. A summary of the legal status of stem cell research, as of late 2004, in the United States (both nationally and within certain states), Australia, Switzerland, and the United Kingdom.

"GRC State Stem Cell Research News and Alerts." Grassroots Connection. Available online. URL: http://grassrootsconnection.com/state_stem_cell_resources.htm. Downloaded on June 11, 2005. A web site run by a group of Parkinson's disease advocacy groups with up-to-date information on the legal status of stem cell research in the U.S. Congress and a number of individual states.

HumGen. "GenBiblio." Available online. URL: http://www.humgen.umontreal.ca/int/GB.cfm. An extensive bibliography of print and electronic references on the political and legal status of stem cell research in all countries of the world.

Jones, Phillip B. C. "Funding of Human Stem Cell Research by the United States," Electronic Journal of Biotechnology, vol. 3, no. 1. Available online. URL: http://www.bioline.org.br/request?ej00004. Posted April 2000. A review of the legal, regulatory, and economic status of stem cell research in the United States with recommendations for the support of such research by federal funds.

Mitchell, Steve. "Stem Cell Lines May Not Meet Guidelines." *Science Daily.* Available online. URL: http://www.sciencedaily.com/upi/index.php?feed=Science&article=UPI-1-20050503-14101600-bc-us-stemcells.xml. Downloaded on June 3, 2005. Guidelines issued by the National Academies of Science may further restrict and complicate research on human embryonic stem cells because some stem cell lines approved for use by President George W. Bush may not meet the new guideline criteria.

Movahed, Shirin. "The Legal Status of Embryos in the United States." Stem-Clone Digest. Available online. URL: http://stemcellsclub.com/SCCC-homesite/stemclonedigest/articlemovahed.html. Accessed November 8, 2005. An excellent review of laws and court cases relating to the legal status of embryos.

Muscati, Sina A. "Defining a New Ethical Standard for Human in Vitro Embryos in the Context of Stem Cell Research." *Duke Law & Technology Review.* Available online. URL: http://www.law.duke.edu/journals/dltr/articles/2002dltr0026.html. Downloaded on May 28, 2005. An iBrief that succinctly reviews the legal issues related to the status of the embryo in stem cell research and suggests that a new, Constitutional definition of the human in vitro embryo be developed and implemented.

The National Academies. "Guidelines Released for Embryonic Stem Cell Research." Available online. URL: http://www4.nationalacademies.org/news.nsf/isbn/0309096537?OpenDocument. Downloaded on May 24, 2005. A press release announcing the forthcoming National Academies book (listed in the Books section) on guidelines for stem cell research in the United States.

National Institutes of Health. "Policy & Guidelines." Available online. URL: http://stemcells.nih.gov/policy/guidelines.asp. Downloaded on May 23, 2005. A compilation of all relevant rules and regulations on stem cell research in the United States, including issues such as funding guidelines, frequently asked questions about government policy and regulations, international shipments of stem cells used for research, and the approval process for research that makes use of stem cells. Documents listed on this site date from December 1, 1999, the earliest of such documents, to April 10, 2002, the latest date for any document of this kind.

Rogers, Adam. "A Stem Cell Confederacy." Wired.com. Available online. URL: http://www.wired.com/wired/archive/12.11/view.html?pg=2. Posted in November 2004. An opinion piece that argues that, while the federal government continues to debate its role in stem cell research, individual states can and should move ahead with their own funding programs for this field of research.

State of California. "Proposition 71: Stem Cell Research. Funding. Bonds. Initiative Constitutional Amendment and Statute." http://www.smartvoter.org/2004/11/02/ca/state/prop/71/. Posted on December 15, 2004. The official statement of Proposition 71, which appeared on the ballot in November 2004, produced by the state's attorney general. The initiative would have created a "California Institute for Regenerative Medicine" to "regulate stem cell research and provide funding, through grants and loans, for such research and research facilities." The proposition passed 59 percent to 41 percent.

United States Department of Health and Human Services. "Fact Sheet: Embryonic Stem Cell Research." Available online. URL: http://www.hhs.gov/news/press/2004pres/20040714b.html. Downloaded on June 1, 2005. A press release reviewing the U.S. government's current policies on stem cell research, based on President George W. Bush's statement of August 12, 2001.

———. "Fact Sheet: Human Pluripotent Stem Cell Research Guidelines." Available online. URL: http://www.hhs.gov/news/press/2001pres/01fsstemcell.html. Posted on January 19, 2001. A press release that summarizes and discusses the guidelines for stem cell research in the United States issued by the National Institutes of Health on August 23, 2000.

University of Minnesota Medical School. "Stem Cell Policy: World Stem Cell Map." Available online. URL: http://mbbnet.umn.edu/scmap.html. Downloaded on May 14, 2005. A map that shows the legal and regulatory status of stem cell research in nations around the world with an extensive and detailed list of references from which the map was drawn. Arguably the best single source of information on the international legal status of SCR currently available.

PRO AND CON ARGUMENTS

BOOKS

Brannigan Michael C, ed. *Ethical Issues in Human Cloning: Cross-disciplinary Perspectives.* New York: Chatham House, 2001. Essays by specialists in the field on the scientific, religious, philosophical, and legal implications of cloning research.

Dewar, Elaine. *The Second Tree: An Investigation into Stem Cells, Cloning, and the Quests for Immortality.* New York: Carroll & Graf, 2005. The author describes her efforts to understand the science and ethics of stem cell research through interviews with important figures in the field in Canada. A reviewer from *Publishers Weekly* praised the book but suggests that "she falters by giving herself far too great a presence, endlessly discussing her scientific ignorance and explaining how she's come to ask the questions she's posing" and that "many American readers may find the extended focus on Canadian politics too narrow."

Green, Ronald M. *The Human Embryo Research Debates: Bioethics in the Vortex of Controversy.* New York: Oxford University Press, 2001. The author, a member of the National Institutes of Health's Human Embryo Research Panel, reflects on his experiences with that group and the ethical issues raised by the relative benefits to be gained from stem cell research compared to the ethical issues that it creates.

Holland, Suzanne, Lebacqz, Karen, and Zoloth, Laurie, eds. *The Human Embryonic Stem Cell Debate.* Cambridge: MIT Press, 2001. A reprint of the 1991 report by the National Bioethics Advisory Commission, *Ethical Issues in Human Stem Cell Research* (see below).

Humber, James M., and Robert F. Almeder, eds. *Stem Cell Research.* Totowa, N.J.: Humana Press, 2003. Seven articles on ethical issues associated with stem cell research, including the rights of human embryos, regulation of research practices, the commercialization of human tissue, and possible effects of embryonic stem cell research on women and minority groups.

Jones, D. Gateth, and Mary Byrne, eds. *Stem Cell Research and Cloning: Contemporary Challenges to Our Humanity.* Hindmarsh, Australia: ATF Press, 2005. Contributors review in five essays the current status of stem cell research, the ethical issues related to such research, and an argument as to the reasons that Christians should not object to the destruction of embryos and permit such research to go forward.

Kristol, William, ed. *The Future Is Now: America Confronts the New Genetics.* Lanham, Md.: Rowman & Littlefield, 2002. A leading spokesperson for the political right wing in the United States assembles a reader that attempts to present all sides in discussions over human cloning, genetic engineering, stem cell research, biotechnology, and human nature with the goal of finding a way to "set moral limits on biological 'progress' before it is too late."

Lauritzen, Paul, ed. *Cloning and the Future of Human Embryo Research.* New York: Oxford University Press, 2001. An anthology of articles on the moral status of the human embryo, the current debate over research on cloning, and problems of developing public policy over stem cell research.

Macintosh, Kerry Lynn. *Illegal Beings: Human Clones and the Law.* New York: Cambridge University Press, 2005. The author provides a review of the pros and cons of human reproductive cloning and then presents a strikingly new and interesting argument, that objections to the practice are "false or exaggerated, inspiring laws that stigmatize human clones as subhuman and unworthy of existence." Harvard University law professor Lawrence Tribe described the book as a "thought-provoking contribution to a constitutional conversation that is just beginning."

McCarthy, Anthony. *Cloning and Stem Cell Research.* London: Linacre Centre for Health Care Ethics, 2004. An analysis of the ethical issues related to cloning and stem cell research from the perspective of a Roman Catholic thinker.

Ruse, Michael, and Christopher Pynes, eds. *The Stem Cell Controversy: Debating the Issues.* Amherst, N.Y.: Prometheus Books, 2003. Twenty-five essays reprinted from other sources are grouped into five sections on "The Science of Stem Cells," "Medical Cures and Promises," "Moral Issues," Religious Issues," and "Policy Issues." A number of articles are taken from publications of the National Commission ("Human Stem Cell Research and the Potential for Clinical Application" and "Ethical Issues in Human Stem Cell Research: Conclusions and Recommendations"), as well as testimony by a number of authorities from various religious groups.

Steinbock, Bonnie. *Life Before Birth: The Moral and Legal Status of Embryos and Fetuses.* New York: Oxford University Press, 1996. A discussion of the rights of embryos and fetuses, offered primarily within the context of the debates over abortion and research on embryos and fetuses.

Waters, Brent, and , Donald Cole-Turner, eds. *God and the Embryo: Religious Voices on Stem Cells and Cloning.* Washington, D.C.: Georgetown University Press, 2003. A series of 11 essays arranged into three sections, on "Frameworks," "Embryos," and "Research," from the perspectives of Protestant, Roman Catholic, Jewish, and Orthodox thinkers.

ARTICLES

Annas, George, Arthur Caplan, and Sherman Elias. "Stem Cell Politics, Ethics, and Medical Progress," *Nature Medicine*, vol. 5, no. 12, December 1999, pp. 1,339–1,341. The authors argue that stem cell research will receive greater support from the general public in the United States if there is a more extended debate on ethical justification for such research.

Bayliss, Françoise. "Our Cells/Ourselves: Creating Human Embryos for Stem Cell Research," *Women's Health Issues*, vol. 10, no. 3, May/June 2000, pp. 140–145. Also available online. URL: http://bioethics.medicine.dal.ca/PubsBaylis/WHI_1.pdf. The author compares two major reports on stem cell research, one issued by the National Bioethics Advisory Commission in 1999 and one by the Human Embryo Research Panel of the National Institutes of Health in 1994 and finds that the former deals more overtly and specifically with ethical issues in using human embryos for stem cell research. She argues that such research should not be encouraged because of the potential legal, psychological, and ethical risks it poses for women.

Blackburn, Elizabeth. "Bioethics and the Political Distortion of Biomedical Science," *New England Journal of Medicine*, vol. 350, no. 14, April 1, 2004, pp. 1,379–1,380. A strong statement of a former member of the President's Council on Bioethics expressing concern that scientific information is being influenced by political factors. She expresses the view that "[t]here is a growing sense that scientific research . . . is being manipulated for political ends."

Blackburn, Elizabeth, and Janet Rowley. "Reason as Our Guide." *PLoS Biology*, vol. 2, no. 4, April 2004, p. e116. Two former members of the President's Council on Bioethics raise questions about the scientific validity of reports recently issued by the council. The article prompted a number of responses, which can be found on the *PLoS Biology* web site at http://www.pubmedcentral.gov/articlerender.fcgi?artid=359389.

Boer, G. J. "Ethical Issues in Neurografting of Human Embryonic Cells," *Theoretical Medicine and Bioethics*, vol. 20, no. 5, September 1999, pp. 461–475. The author points out that human embryos have been used in scientific research "for decades" with little or no ethical issues having been raised. Now that that situation has changed, he suggests that decisions about the transplantation of embryonic stem cells into the brain must be decided on the basis of the risks and benefits for both donor and acceptor (the patient), and that, in certain cases, such research and/or treatment should be permitted.

Borge O. J., and K. Evers. "Aspects on Properties, Use and Ethical Considerations of Embryonic Stem Cells—A Short Review," *Cytotechnology*, vol. 41, nos. 2–3, January 2003, pp. 59–68. As the title suggests, a review of the current status of the science, technology, potential applications, and ethical issues related to stem cell research.

Bortolotti, Lisa, and John Harris. "Stem Cell Research, Personhood and Sentience," *Reproductive Biomedicine Online*, vol. 10, suppl. 1, March 2005, pp. 68–75. The authors argue that "[e]arly human embryos do not satisfy the requirements for personhood," thus negating an important argument against the use of embryonic stem cells in research. The article is included in a special supplement on "Ethics, Science and Moral Philosophy of Assisted Human Reproduction," a collection of papers presented at the

First International Conference on the Ethics, Science and Moral Philosophy of Assisted Human Reproduction, held in London on September 30–October 1, 2004.

Branick, V., and M. T. Lysaught. "Stem Cell Research: Licit or Complicit? Is a Medical Breakthrough Based on Embryonic and Fetal Tissue Compatible with Catholic Teaching?" *Health Progress*, vol. 80, no. 5, September–October 1999, pp. 37–42. An article prompted by the National Institutes of Health's decision to fund embryonic stem cell research and asking whether Roman Catholic universities can support that decision and sponsor SCR. The authors conclude that the answer is "probably not" as such research conflicts with Catholic teachings.

Bruck, Connie. "On the Ballot—Hollywood Science: Should a Ballot Initiative Determine the Fate of Stem-Cell Research?" *The New Yorker*, October 18, 2004, pp. 62–82. A good review of the history of stem cell research and controlling legislation and regulation in the United States.

Cahill, Lisa Sowle. "Social Ethics of Embryo and Stem Cell Research," *Women's Health Issues*, vol. 10, no. 3, May–June 2000, pp. 131–135. Most ethical debates over embryonic stem cell research focus on the status of the embryo as the deciding factor as to whether such research should be permitted, but other factors are important also, such as the economic impetus that drives much SCR. Those factors need to be taken into account in determining whether SCR should be allowed and/or funded.

Charo, Alta. "The Hunting of the Snark: The Moral Status of Embryos, Right-to-Lifers, and Third World Women," *Stanford Law and Policy*, vol. 6, no. 2, 1995, pp. 11–27. The author considers the question how, if at all, a governmental body can make decisions as to which actions are moral and which are not, in this case, on the moral status of the embryo. She concludes that the 1994 report of the NIH Human Embryo Research Panel was unable and even unwilling to resolve this issue and, as a result, made recommendations that were unduly restrictive on the conduct of stem cell research.

Cheshire, William P., Jr., et al. "Stem Cell Research: Why Medicine Should Reject Human Cloning," *Mayo Clinic Proceedings*, vol. 78, no. 8, August 2003, pp. 1,010–1,018. Also available online. URL: http://courses.washington.edu/devneuro/week9pdfs/MayoClinic.pdf. The authors explain why they are opposed to all forms of human cloning, whether it be for reproductive purposes or for the purpose of producing embryos for stem cell research. Such research, they say, "represents an abuse of scientific freedom, not its realization."

Childress, James F. "Sources of Stem Cells: Ethical Controversies and Policy Developments in the United States," *Fetal Diagnosis and Therapy*, vol. 19, no. 2, March–April 2004, pp. 119–123. The author reviews the history of the ethical debate over stem cell research in the United States and concludes that fundamental disagreements about ethical issues ensure that this debate will not be resolved, but will remain a constant factor in SCR.

Coors, Marilyn E. "Therapeutic Cloning: From Consequences to Contradiction," *Journal of Medicine and Philosophy*, vol. 27, no. 3, June 2002, pp. 297–317. A commentary on the British parliament's decision to approve embryonic stem cell research, which the author calls "the active and unprecedented government support for the generation and destruction of human embryonic life merely as a means of medical advancement." She concludes that the action reduces the value of human life to its utilitarian value, neglecting other factors that should be taken into consideration.

Curtis, Michele G. "Cloning and Stem Cells: Processes, Politics, and Policy," *Current Women's Health Reports*, vol. 3, no. 6, December 2003, pp. 492–500. Also available online. URL: http://www.biomedcentral.com/content/pdf/cr-wr3632.pdf. The development of science policy depends on an up-to-date and accurate understanding of the state of technological development in the field, but the politicization of science in the United States in recent years makes it "almost impossible" to develop a sound national policy on stem cell research.

Daar, Abdallah S., et al. "Stem Cell Research and Transplantation: Science Leading Ethics," *Transplantation Proceedings*, vol. 36, no. 8, pp. 2,504–2,506. An overview of the ethical problems related to stem cell research.

Davis, Dena S. "Informed Consent for Stem Cell Research in the Public Sector," *Journal of the American Medical Women's Association*, vol. 55, no. 5, Fall 2000, pp. 270–274. A number of issues are involved in asking men and women to donate spare embryos for research, but if careful thought is given to these issues, there is no reason that true informed consent cannot be obtained for the use of such embryos in stem cell research.

De Castro, Leonard D. "Exploitation in the Use of Human Subjects for Medical Experimentation: A Reexamination of Basic Issues," *Bioethics*, vol. 9, no. 3–4, July 1995, pp. 259–268. The author raises questions of the effects of power imbalance—for example, between highly educated researchers and often poorly educated volunteers—in obtaining "informed consent" on experiments involving the donation of medical products.

De Volder, Katrien. "Creating and Sacrificing Embryos for Stem Cells," *Journal of Medical Ethics*, vol. 31, no. 6, June 2005, pp. 366–370. The distinction between creating embryos for in vitro fertilization and for stem cell research is very narrow, and anyone willing to accept the former procedure is philosophically obligated to accept the second as well.

De Wert, Guido, and Christine Mummery. "Human Embryonic Stem Cells: Research, Ethics and Policy," *Human Reproduction*, vol. 18, no. 4, April 2003, pp. 672–682. An overview of the technology of stem cell research and the major ethical issues that surround the technology involved.

Dinc, Leyla. "Ethical Issues Regarding Human Cloning: A Nursing Perspective," *Nursing Ethics*, vol. 10, no. 3, May 2003, pp. 238–254. A review of the technology and possible applications of stem cell research with an analysis of the ethical issues involved from a nursing standpoint.

Doerflinger, Richard. "Destructive Stem Cell Research on Human Embryos," *Origins*, vol. 28, April 29, 1999, pp. 770–773. The author argues that the use of embryonic stem cells in research is ethically improper and that, in any case, recent advances in research have shown that adult stem cells may be able to meet all the research needs for therapeutic applications claimed for embryonic stem cells.

———. "The Ethics of Funding Embryonic Stem Cell Research: A Catholic Viewpoint," *Kennedy Institute of Ethics Journal*, vol. 9, no. 2, June 1999, pp. 137–150. The author, deputy director of the secretariat for pro-life activities of the U.S. Conference of Catholic Bishops, makes two fundamental arguments against stem cell research: first, that "human life is a continuum from the one-cell stage onward," and, therefore, any human embryo, at any stage of development, must be treated as one would treat any human being; and second, that all human beings, including human embryos, must always be treated "as an end in himself or herself; not merely as a means to other ends."

Dorff, Elliot N. "Stem Cell Research," *Conservative Judaism Journal*, vol. 55, no. 3, Spring 2003, pp. 3–29. A review of the ethical issues raised by stem cell research and the position of conservative Jewish thought on this question. The author explains why SCR is regarded as an acceptable procedure within this religious tradition.

Drazen, Jeffrey M. "Embryonic Stem-Cell Research—The Case for Federal Funding," *New England Journal of Medicine*, vol. 351, no. 17, October 21, 2004, pp. 1,789–1,790. The author notes that governmental funding of embryonic stem cell research has now been approved in many developed nations and the failure of the U.S. government to follow suit is likely to result in a serious "brain drain" of researchers in the field in the near future.

Edwards, R. G. "Ethics and Moral Philosophy in the Initiation of IVF, Preimplantation Diagnosis and Stem Cells," *Reproductive Biomedicine Online*, vol. 10, suppl. 1, March 2005, pp. 1–8. A general overview of the moral issues that have been raised by developments in the field of stem cell research. The paper is an introduction to a special supplement on "Ethics, Science and Moral Philosophy of Assisted Human Reproduction," a collection of papers presented at the First International Conference on the Ethics, Science and Moral Philosophy of Assisted Human Reproduction, held in London on September 30–October 1, 2004.

Faden, R. R., et al. "Public Stem Cell Banks: Considerations of Justice in Stem Cell Research and Therapy," *Hastings Center Report*, vol. 33, no. 6, November–December 2003, pp. 13–27. Among the many ethical issues surrounding the use of stem cell research is the question as to how one provides equal access to the fruits of such research. The authors suggest that care must be taken to ensure that the benefits of SCR do not accrue unproportionately to white middle-class citizens.

Farley, Margaret. "Roman Catholic Views on Research Involving Human Embryonic Stem Cells," in Suzanne Holland, Karen Lebacqz, and Laurie Zoloth, eds. *The Human Embryonic Stem Cell Debate*. Cambridge: MIT Press, 2001, pp. 113–118. An article based on a paper presented by the author to the National Bioethics Advisory Commission in 1999 that argues that positions other than those espoused by the Vatican and most Catholic theologians can be developed to support the use of human embryonic stem cells for research purposes.

Fins, Joseph Jack, and Madelaine Schachter. "Patently Controversial: Markets, Morals and the President's Proposal for Embryonic Stem Cell Research," *Kennedy Institute of Ethics Journal*, vol. 12, no. 3, September 2002, pp. 265–278. An analysis of the effects of President George W. Bush's policy on stem cell research, offered from the standpoint of patent law, privacy, and informed consent.

Fischbach, Gerald D., and Ruth L. Fischbach. "Stem Cells: Science, Policy, and Ethics," *Journal of Clinical Investigation*, vol. 114, no. 10, November 2004, pp. 1,364–1,370. A discussion of the promise provided by stem cell research, the research challenges that need to be answered, and ethical issues that have arisen as a result of progress in SCR.

FitzPatrick, William. "Surplus Embryos, Nonreproductive Cloning, and the Intend/Foresee Distinction." *Hastings Center Report*, vol. 33, no. 3, May 2003, pp. 29–36. The author argues that "There is . . . a real moral difference between creating embryos expressly for medical research and conducting research on embryos that are left over from infertility treatments. To create an embryo intending all along to destroy it is worse. But in the end, it isn't so much worse that we should ban all nonreproductive cloning."

Gilbert, David M. "The Future of Human Embryonic Stem Cell Research: Addressing Ethical Conflict with Responsible Scientific Research," *Medical Science Monitor*, vol. 10, no. 5, May 2004, pp. RA99–103. Also available online. URL: http://www.medscimonit.com/pub/vol_10/no_5/4448.pdf. An exploration of various levels of ethical issues that may arise as a result of progress in stem cell research.

Gordijn Bert. "The Troublesome Concept of the Person," *Theoretical Medicine and Bioethics*, vol. 20, no. 4, August 1999, pp. 347–359. Introducing the concept of "personhood" into discussions about stem cell research is philosophically and ethically erroneous and only serves to further the confusion of the issues involved. The author suggests an alternative way of debating the ethical issues involved in SCR that does not involve the use of the concept.

Gross, Michael. "Swiss Back Stem-Cell Studies," *Current Biology*, vol. 15, no. 2, January 26, 2005, p. R35. The first national referendum in the world results in a favorable vote in support of stem cell research in Switzerland.

Guenin, Louis M. "A Failed Noncomplicity Scheme," *Stem Cells and Development*, vol. 13, no. 5, October 2004, pp. 456–459. An argument that the U.S. government's policy on stem cell research does not avoid the ethical problems it was designed to bypass, and that the position really depends on scientific progress rather than religious or ethical concerns.

Harris, John. "The Concept of the Person and the Value of Life," *Kennedy Institute of Ethics Journal*, vol. 9, no. 4, December 1999, pp. 293–308. Harris, a member of the Centre for Social Ethics and Policy at the University of Manchester, England, points out the importance of the concept of "personhood" in debates over SCR and attempts to develop a clear definition of that term for use in such debates.

Harris, John, ed. "Stem Cell Research," *Bioethics*, vol. 16, no. 6, November 2002. A special issue on stem cell research including articles on topics such as "Going to the Roots of the Stem Cell Controversy," "The Embryonic Stem Cell Lottery and the Cannibalization of Human Beings," "Principles of Ethical Decision Making Regarding Embrionic *[sic]* Stem Cell Research in Germany," "Benefiting from 'Evil': An Incipient Moral Problem in Human Stem Cell Research," and "Embryonic Stem Cell Research and Therapy: The Need for a Common European Legal Framework."

Helig, Steve. "Stem Cell Science and Politics: A Talk with Elizabeth Blackburn, Ph.D.," *San Francisco Medicine*, vol. 77, no. 9, October 2004. Also available online. URL: http://www.sfms.org/sfm/sfm1004b.htm. A fascinating interview with a biomedical researcher who was appointed to President George W. Bush's Council on Bioethics in 2001 about her experience in working on that council until she was "fired" by the president in 2004. Also see Blackburn, Elizabeth. "Bioethics and the Political Distortion of Biomedical Science," in the Articles section.

Heyd, David. "Experimenting with Embryos: Can Philosophy Help?" *Bioethics*, vol. 10, no. 4, October 1996, pp. 292–309. The author points out ways in which the discipline of philosophy can help clarify and contribute to the debate over the use of human embryos in stem cell research.

Holm, Søren."Going to the Roots of the Stem Cell Controversy, *Bioethics*, vol. 16, no. 6, November 2002, pp. 493–507. A review of the scientific and ethical bases of the stem cell research debate that considers stem cells and the status of the embryo, women as the sources of ova for stem cell production, the use of ova from other species, so-called slippery slope problems toward reproductive cloning, the public presentation of stem cell research, and the evaluation of scientific uncertainty and its implications for public policy.

———. "The Spare Embryo: a Red Herring in the Embryo Experimentation Debate," *Health Care Analysis*, vol. 1, no. 1, June 1993, pp. 63–66. An interesting argument that "the question of whether it is preferable to use spare or specifically produced ('research') embryos for destructive embryo experimentation . . . is morally uninteresting, but rhetorically useful for both sides in the debate."

"Human Primordial Stem Cells: A Symposium," *Hastings Center Report*, vol. 29, no. 2, March–April 1999, pp. 30–48. A group of articles prepared in response to an invitation by the Geron Corporation to solicit expert opinion on ethical issues involved in stem cell research.

Josko, R. M. "Clones, Parents, Persons and the Law in Australia," *Clinica Terapeutica*, vol. 153, no. 6, November–December 2002, pp. 421–427. A survey of laws in Great Britain and Australia with regard to cloning and artificial reproduction finds, according to the author, that those laws are "in contradiction with fundamental concepts upon which our society is built" and "bound to fail."

Juengst, Eric, and Michael Fossel. "The Ethics of Embryonic Stem Cells—Now and Forever, Cells Without End," *JAMA*, vol. 284, no. 24, December 27, 2000, pp. 3,180–3,184. The ethical issues surrounding stem cell research are far more complex than many commentators are willing to accept and go beyond purely scientific or political considerations.

Kalb, Claudia, and Debra Rosenberg. "Nancy's Next Campaign. The Former First Lady's Passion for Stem-cell Research Has Fueled a Political Battle. Where Does the Science Stand?" *Newsweek*, June 21, 2004, pp. 38–44. A discussion of former First Lady Nancy Reagan's efforts to promote stem cell research in the United States, motivated largely by her husband's long battle with Alzheimer's disease, with a review of the current status of SCR.

Kamm, F. M. "Embryonic Stem Cell Research: A Moral Defense," *Boston Review*, vol. 27, no. 5, November 2002. Also available online. URL: http://www.bostonreview.net/BR27.5/kamm.html. The author argues that embryonic stem cell research can be justified morally because destruction of the embryo is not bad for the embryo itself. The benefits that are to be expected from such research fully justify its conduct.

Kass, Leon R. "L'chaim and its Limits: Why Not Immortality?" *First Things: A Monthly Journal of Religion and Public Life*, vol. 113, May 2001, pp. 17–24. An essay on some problems in modern medical bioethics, including the problems surrounding stem cell research, from the standpoint of traditional Jewish religious thought.

Keown, J. "The Polkinghorne Report on Fetal Research: Nice Recommendations, Shame about the Reasoning," *Journal of Medical Ethics*, vol. 19, no. 2, June 1993, pp. 114–120. In 1989, a committee, headed by the Reverend Dr. John Polkinghorne, was appointed to consider a revision of the U.K.'s Code of Practice as a result of the first experiments on the transplantation of fetal tissue into the brains of individuals with Parkinson's Disease. This article claims to be the first to subject the Polkinghorne Report to "sustained ethical and legal scrutiny" and finds that the report produced a satisfactory revision of the code, but was seriously lacking in some of the committee's moral and ethical reasoning.

Kitzingera, Jenny, and Clare Williams. "Forecasting Science Futures: Legitimising Hope and Calming Fears in the Embryo Stem Cell Debate," *Social Science & Medicine*, vol. 61, no. 3, August 2005, pp. 731–740. The authors point out that many socioscientific issues are debated not so much on the current status of science as on potential future outcomes, outcomes that may lie many years or decades in the future. They discuss ways in which this tendency has influenced the debate over the use of embryonic stem cells in research and the increased interaction between scientific expertise and emotional attitudes toward contentious issues in science.

Koehn, Daryl. "The Ethics of Biobusiness, Technology, and Genetic Engineering," *Bulletin of Science, Technology & Society*, vol. 20, no. 1, January 2000, pp. 10–18. The author argues that essentially all of the ethical discussions of modern medical technology, such as stem cell research, "completely [miss] the larger ethical issues arising from the fact that our practices, including medicine, have become thoroughly technological in character" and need a new context within which to be analyzed. The author proposes to outline some features of that new context.

Larijani, Bagher, and Farzaneh Zahedi. "Islamic Perspective on Human Cloning and Stem Cell Research," *Transplantation Proceedings*, vol. 36, no. 10, December 2004, pp. 3,188–3,189. An analysis of the religious issues involved in the stem cell research debate from the standpoint of the Islamic religion and teachings in the Qu'ran.

Lillge, Wolfgang. "The Case for Adult Stem Cell Research," *21st Century Science and Technology*, vol. 14, no. 2, Winter, pp. 2,001–2,002. Also available online. URL: http://www.21stcenturysciencetech.com/articles/winter01/stem_cell.html. The author argues that inadequate attention is being paid to the possibilities of using adult stem cells in research to develop medical therapies, placing an undue influence on the possible role of embryonic stem cells in such research.

Lo, Bernard, et al. "Informed Consent in Human Oocyte, Embryo, and Embryonic Stem Cell Research," *Fertility and Sterility*, vol. 82, no. 3, September 2004, pp. 559–563. A key element in research involving human subjects and materials obtained from them is getting their "informed consent" to participate in such studies and/or to use such materials. The authors discuss the ethical and practical issues involved in making sure that this important element is properly satisfied.

MacDonald, Chris. "Stem Cell Ethics and the Forgotten Corporate Context," *American Journal of Bioethics*, vol. 2, no. 1, Winter 2002, pp. 54–56. Most discussions of the ethical issues surrounding stem cell research ignore the significant impact of corporate interests in the development of this technology. The author analyzes three articles that make at least some attempt to include business ethics into their discussions and suggests ways in which corporate policy can and should have an even more important role in such discussions.

Mahowald, Mary Briody. "The President's Council on Bioethics, 2002–2004," *Perspectives in Biology and Medicine*, vol. 48, no. 2, Spring 2005, pp. 159–171. The author reviews the history of the President's Council on Bioethics, reviews its published reports, discusses key definitions adopted by the council, examines recommendations made by the council, and analyzes some of the viewpoints held by members of the council who contributed to development of its reports. Also of interest in the issue are articles by former members of the council Elizabeth Blackburn ("Thoughts of a Former Council Member") and William F. May ("The President's Council on Bioethics: My Take on Some of Its Deliberations").

Marincola, Elizabeth. "Research Advocacy: Why Every Scientist Should Participate," *PLoS Biology*, vol. 1, no. 3, December 2003, e71. A plea for a much greater participation on the part of working scientists in the formation of science policy in the United States.

Marshall, Eliot. "The Business of Stem Cells," *Science*, vol. 287, no. 5457, February 25, 2000, pp. 1,419–1,421. The author considers some of the business implications of the promising new world of stem cell research. The article is particularly interesting when read in conjunction with a later article with the same title written four years later after the administration of President George W. Bush prohibited federal funding on most types of stem cell research (see Debora Spar, "The Business of Stem Cells," *New England Journal of Medicine*, vol. 351, no. 3, July 15, 2004, pp. 211–213.)

McCormick, Richard A. "Who or What Is the Preembryo?" *Kennedy Institute of Ethics Journal*, vol. 1, no. 1, March 1991, pp. 1–15. The author, a

Jesuit monk, struggles with the question as to whether a "pre-embryo" is actually a person or not, if and when it attains "personhood," and what moral value it has. He concludes that enough doubt exists about the answers to these questions that the destruction of pre-embryos for possible therapeutic purposes is not justified.

McGee, G., and A. Caplan, "The Ethics and Politics of Small Sacrifices in Stem Cell Research," *Kennedy Institute of Ethics Journal*, vol. 9, no. 2, June 1999, pp. 151–158. The authors take a strong stand in support of stem cell research, arguing that the potential medical benefits are so great that the sacrifice of embryos to obtain stem cells is easily justified, such that "the moral imperative of compassion . . . compels stem cell research."

McMahan, Jeff. "Cloning, Killing, and Identity," *Journal of Medical Ethics*, vol. 25, no. 2, April 1999, pp. 77–86. The author argues that the embryo/fetus/unborn human does not begin to resemble "you or me" until the seventh month of gestation, so that the killing of such an organism before that time is justified if it results in producing some medical benefit to other humans.

Meilaender, Gilbert. "The Point of a Ban: Or, How to Think about Stem Cell Research," *Hastings Center Report*, vol. 31, no. 1, January–February 2001, pp. 9–16. The author approaches the issue of stem cell research from a general moral argument about the value of human life, concluding that potential life (embryos) should not be sacrificed for some possible future therapeutic benefits.

Mummery, Christine. "Stem Cell Research: Immortality or a Healthy Old Age?" *European Journal of Endocrinology*, vol. 151, suppl. 3, November 2004, pp. U7–U12. A review of the state-of-the-art technology in SCR as a basis for determining whether such research is ethical or not and, if so, under what circumstances.

Nisbet, Matthew C., Dominique Brossard, and Adrianne Kroepsch. "Framing Science: The Stem Cell Controversy in an Age of Press/Politics," *The Harvard International Journal of Press/Politics*, vol. 8, no. 2, April 2003, pp. 36–70. Report of a study that attempts to determine the role of the mass media in the evolution of the stem cell controversy by analyzing articles in the *New York Times* and *The Washington Post* from 1975 through 2001. The authors attempt to discover the forces that cause a particular scientific issue to gain, maintain, and/or lose public attention.

Novak, Michael. "The Stem Cell Slide: Be Alert to the Beginnings of Evil," *National Review*, vol. 53, no. 17, September 3, 2001, pp. 17–18. A strong statement of opposition to any type of research in which an embryo, pre-embryo, or similar structure is destroyed for research purpose. The author argues that such research is driven by desire for economic profit and personal advancement rather than concerns for human welfare.

O'Hara, N. "Ethical Consideration of Experimentation Using Living Human Embryos: The Catholic Church's Position on Human Embryonic Stem Cell Research and Human Cloning," *Clinical and Experimental Obstetrics & Gynecology*, vol. 30, no. 2–3, 2003, pp. 77–81. While acknowledging the very significant clinical benefits that may accrue from stem cell research, the author points out that the Catholic Church teaches that life begins at the moment of conception and that the destruction of embryos for medical research—or for any other reason—is, therefore, immoral and unacceptable.

O'Rourke, Kevin D. "Stem Cell Research—Prospects and Problems," *National Catholic Bioethics Quarterly*, vol. 4, no. 2, Summer 2004, pp. 289–299. A review of the conventional Roman Catholic view on embryonic stem cell research, pointing out that such research depends on the destruction of a human life, the embryo, which is comparable to abortion and, therefore, forbidden. The author argues for greater attention to the possible uses of adult stem cells in research.

Robertson, John A. "Ethics and Policy in Embryonic Stem Cell Research," *Kennedy Institute of Ethics Journal*, vol. 9, no. 2, June 1999, pp. 109–136. The author reviews other controversies over embryo research and concludes that stem cell research using donated spare embryos or embryos produced specifically for SCR is ethically acceptable and worthy of federal funding.

———. "Human Embryonic Stem Cell Research: Ethical and Legal Issues," *Nature Reviews Genetics*, vol. 2, no. 1, January 2001, pp. 74–78. The author discusses the ethical and legal arrangements that must be addressed in order for research on human embryonic tissue to occur "while maintaining respect for human life generally."

Romano, Gaetano. "Stem Cell Transplantation Therapy: Controversy over Ethical Issues and Clinical Relevance," *Drug News & Perspectives*, vol. 17, no. 10, December 2004, pp. 637–645. A review of the scientific potential of stem cells in therapeutic research and the ethical issues that must be resolved before research in this field can go forward. The author suggests that one key to the resolution of this problem is a greatly improved understanding of the nature and function of stem cells.

Ryan, Kenneth J. "The Politics and Ethics of Human Embryo and Stem Cell Research," *Women's Health Issues*, vol. 10, no. 3, May–June 2000, pp. 105–110. The author claims that stem cell research "so greatly affect[s] women's reproductive interests and the promise of health benefits" that it should not be "held hostage" to pro-life interests in the current Congress and deserves to receive governmental sponsorship.

Sandel, Michael J. "Embryo Ethics—The Moral Logic of Stem-Cell Research," and Paul R. McHugh, "Zygote and 'Clonote'—The Ethical Use of Embryonic Stem Cells," *New England Journal of Medicine*, vol. 351, no. 3, July 15, 2004, pp. 207–209. The journal asked two members of the President's Council on Bioethics to comment on questions about the federal funding of stem cell research. These two articles provide the responses given by those two individuals. Also see responses and reactions to these articles in vol. 351, no. 16, October 14, 2004, pp. 1,687–1,690.

Sarfati, Jonathan. "Stem Cells and Genesis," *Creation Magazine*, vol. 15, no. 3, December 2001, pp. 19–26. The author claims that adult stem cell research can provide the information and discoveries needed to deal with most human medical problems and that the Bible and Christian doctrine prohibit the use of embryonic stem cells for research because they require the taking of a human life in order for those cells to be harvested.

Savulescu, Julian. "The Embryonic Stem Cell Lottery and the Cannibalization of Human Beings," *Bioethics*, nol. 16, no. 6, November 2002, pp. 508–529. Whether or not the killing of a person (embryo) is justified depends on the "Embryonic Stem Cell Lottery," which depends on "(1) whether innocent people at risk of being killed for ES cell research also stand to benefit from the research and (2) whether their overall chances of living are higher in a world in which killing and ES cell research is conducted."

Savulescu, Julian, and John Harris. "The Great Debates," *Cambridge Quarterly of Healthcare Ethics*, vol. 13, no. 1, January 2004, pp. 68–96. An exchange of viewpoints between two world-famous philosophers on the issue of stem cell research, beginning with Harris's paper "Stem Cells, Sex, and Procreation" (from this journal, vol. 12, no. 4), followed by a rejoinder by Savulescu, a response from Harris, and a final paper written by the two together on points of common agreement.

Shannon, Thomas A. "Human Embryonic Stem-cell Therapy," *Theological Studies*, vol. 62, no. 4, December 2001, pp. 811–824. A review of the scientific issues related to stem cell research, competing moral and ethical claims regarding the early embryo, and issues for health care and public policy raised by SCR.

———. "Stem-Cell Research: How Catholic Ethics Guide Us," *Catholic Update*, January 2002, p. 1ff. Also available online. URL: http://www.americancatholic.org/Newsletters/CU/ac0102.asp. An explanation of the way in which Roman Catholic doctrine is relevant to stem cell research.

Shannon, Thomas A., and Allan B. Wolter. "Reflections on the Moral Status of the Pre-Embryo," *Theological Studies*, vol. 51, no. 4, December 1990, pp. 603–626. Two Roman Catholic scholars present a view of the pre-embryo as a developing entity that may not necessarily be subject to the same level of respect that an older embryo, a fetus, or a child that has been born receives, a view that differs from that offered by the Vatican.

Sherley, James L. "Human Embryonic Stem Cell Research: No Way Around a Scientific Bottleneck," *Journal of Biomedicine and Biotechnology*, vol. 2, 2004, pp. 71–72. The author argues that some stem cell researchers who are "attempting to run end-around plays against current US government policies" are causing irreparable damage to the field in general because they are causing a reduction in overall funding for the field and raising hopes for progress that cannot be realized in the foreseeable future.

Sowle-Cahill, Lisa. "Stem Cells: A Bioethical Balancing Act," *America*, vol. 184, no. 10, March 26, 2001, pp. 14–19. Stem cell research raises moral issues not only of the personhood of the embryo and its right to survive, but also more general moral questions of the values of human life in a society in which economic benefits have come to play such an important role in socioscientific issues.

Steinbock, Bonnie. "What Does 'Respect for Embryos' Mean in the Context of Stem Cell Research?" *Women's Health Issues*, vol. 10, no. 3, May–June 2000, pp. 127–130. An attempt to clarify a commonly used phrase in discussion of stem cell research, "respect," and its implications for SCR. The author argues that "respect" for an early embryo is different from "respect" for persons "as autonomous agents." She also argues that all embryos should be treated in the same way, whether they are "spare" embryos or embryos created for research.

"Stem Cell Research and Ethics," *Science*, vol. 287, no. 5457, February 25, 2000, pp. 1,417–1,446. A special section of the magazine devoted to a discussion of the science of stem cell research and its ethical, economic, and legal implications. Some of the articles included in the section deal with stem cells in epithelial tissue, mammalian neural stem cells, the role of patients in the debate over stem cell research, activities in stem cell research in Europe, and a review of some of the economic issues involved in stem cell techonology.

Tesarik, Jan, and Ermanno Greco. "A Zygote Is Not an Embryo: Ethical and Legal Considerations," *Reproductive BioMedicine Online*, vol. 9, no. 1, July 1, 2004, pp. 13–16. The authors discuss the importance of terminology in making legal, ethical, and scientific decisions in the field of stem cell research and conclude that "the application of laws aimed at the protection of early human life may have inadequate consequences for the efficacy of the current techniques of human infertility treatment." They review the biology of embryonic development in an attempt to develop a clearer and more specific definition of early life to help resolve this problem.

Towns, C. R., and D. G. Jones. "Stem Cells, Embryos, and the Environment: A Context for Both Science and Ethics," *Journal of Medical Ethics*, vol. 30, no. 4, August 2004, pp. 410–413. In the debate over stem cell research, ethicists and scientists often talk at cross purposes to each other, or do not talk to each other at all. The authors offer a review of the fundamental scientific principles in SCR around which ethical discussions should be based.

Warnock, Mary. "Do Human Cells Have Rights?" *Bioethics*, vol. 1, no. 1, January 1987, pp. 1–14. The chair of Great Britain's Committee of Inquiry Into Human Fertilisation and Embryology explains her committee's thinking and conclusions on this subject, pointing out that embryos have no legal rights in Great Britain, but that "the public's moral feelings must be taken into account when government policy is formulated."

Yamamoto, Keith R. "Bankrolling Stem-Cell Research with California Dollars," *New England Journal of Medicine*, vol. 351, no. 17, October 21, 2004, pp. 1,711–1,713. An adviser to the committee working for the passage of Proposition 71 in California's November 2004 election explains why states will have to make up for the federal government's unwillingness to support stem cell research.

Zoloth, Laurie. "Reasonable Magic and the Nature of Alchemy: Jewish Reflections on Human Embryonic Stem Cell Research," *Kennedy Institute of Ethics Journal*, vol. 12, no. 1, March 2002, pp. 65–93. Beginning with the tendency of some writers to refer to regenerative medicine as a "magical" process, the author reviews traditional Jewish views about magic and "forbidden knowledge" with regard to their influence on current discussions of the ethics of stem cell research.

Zwanziger, Lee L. "Biology, Ethics, and Public Policy: Deliberations at an Unstable Intersection," *The Anatomical Record Part B: The New Anatomist*, vol. 275B, no. 1, pp. 185–189. The debate over stem cell research involves inputs from three fields: biological theory and practice, ethics, and public policy. The author shows how changes in any one of these fields—such as developments in biological theories or the evolution of new ethical principles—changes the way the debate over stem cell research is phrased and resolved.

REPORTS AND STATEMENTS

American Medical Association. "Embryonic/Pluripotent Stem Cell Research and Funding. Report 15 of the Council on Scientific Affairs." (I-99). Chicago: AMA, December 1999. Available online. URL: http://www.ama-assn.org/ama/pub/category/13594.html. A report that examines the science, policy implications, and AMA initiatives and activities with regard to stem cell research, the positions of those who support and those who oppose SCR, and current AMA policy on this issue.

Committee on Culture, Science and Education. "Human Stem Cell Research" (Document 9902). Strassbourg, France: Parliamentary Assembly of the Council of Europe, September 11, 2003. Also available online. URL: http://assembly.coe.int/Documents/WorkingDocs/doc03/EDOC9902.htm and in print form in *Human Reproduction and Genetics Ethics*, vol. 10, no. 2, 2004, pp. 53–67. A report to the European Council on stem cell research that calls on member states, among other things, to (1) "promote stem cell research as long as it respects the life of human beings in any state of their development," (2) "encourage scientific techniques that are not socially and ethically divisive in order to advance the use of cell pluripotency and develop new methods in regenerative medicine," (3) "ensure that any research on stem cells involving the destruction of human embryos is duly authorised and monitored by the appropriate national bodies," and (4) "give priority to the ethical aspects of research over those of a purely utilitarian and financial nature."

Ethical and Policy Issues in Research Involving Human Participants. Bethesda, Md.: National Bioethics Advisory Commission, August 2001. A report issued in response to President Bill Clinton's charge when he created the NBAC in 1995 to "provide advice and make recommendations to the National Science and Technology Council and to other appropriate government entities regarding the following matters: 1) the appropriateness of departmental, agency, or other governmental programs, policies, assignments, missions, guidelines, and regulations as they relate to bioethical issues arising from research on human biology and behavior; and 2) applications, including the clinical applications, of that research."

Ethics Committee of the American Society for Reproductive Medicine. "Donating Spare Embryos for Embryonic Stem-cell Research." *Fertility and Sterility*, vol. 82, suppl. 1, September 2004, pp. 224–227. A report on the ethical issues involved in the donation of embryos originally produced for some other purpose to stem cell and similar types of research by the group's special committee on ethical questions.

National Bioethics Advisory Commission. *Ethical Issues in Human Stem Cell Research.* Washington, D.C.: National Technical Information Service, U.S. Department of Commerce, September 1999. A three-volume report on ethical issues related to human embryonic stem cell research. The three volumes are: "Report and Recommendations of the National Bioethics Advisory Commission," "Commissioned Papers," and "Religious Perspectives." The report is also available online at http://www.georgetown.edu/research/nrcbl/nbac/pubs.html.

Nuffield Council on Bioethics. "Stem Cell Therapy: The Ethical Issues." London: Nuffield Council on Bioethics, April 2000. A discussion paper that summarizes the result of a roundtable meeting held on September

29, 1999, sponsored by the Nuffield Council on Bioethics, to discuss the key ethical issues raised by stem cell therapy. Participants in the roundtable concluded that "the removal and cultivation of cells from a donated embryo does not indicate lack of respect for the embryo." They concluded, therefore, that "there are no grounds for making a moral distinction between research into diagnostic methods or reproduction which is permitted under UK legislation and research into potential therapies which is not permitted," and recommended that research of the latter type be permitted in the U.K.

Outka, Gene. "The Ethics of Stem Cell Research." A briefing paper prepared for the President's Council on Bioethics meeting of April 2002. Outka offers a detailed summary of the ethical views about SCR expressed from those "on the right," "on the left," and "in the middle."

Pontifical Academy for Life. "Declaration on the Production and the Scientific and Therapeutic Use of Human Embryonic Stem Cells." Available online. URL: http://www.vatican.va/roman_curia/pontifical_academies/acdlife/documents/rc_p_acdlife_doc_20000824_cellule-staminali_en.html. Downloaded on June 1, 2005. A declaration issued on August 25, 2000, explaining why the use of embryonic stem cells in research is not permitted under Roman Catholic canon law, although the use of adult stem cells presents no such ethical problems.

Sugarman, J., et al. "Ethical Issues in Umbilical Cord Blood Banking. Working Group on Ethical Issues in Umbilical Cord Blood Banking," *JAMA*, vol. 278, no. 11, September 17, 1997, pp. 938–943. Report of a committee appointed to study ethical issues involved in the banking of umbilical cord blood for possible later use in stem cell research. The committee concluded that "(1) [u]mbilical cord blood technology is promising although it has several investigational aspects; (2) during this investigational phase, secure linkage should be maintained of stored UCB [umbilical cord blood] to the identity of the donor; (3) UCB banking for autologous use is associated with even greater uncertainty than banking for allogeneic use; (4) marketing practices for UCB banking in the private sector need close attention; (5) more data are needed to ensure that recruitment for banking and use of UCB are equitable; and (6) the process of obtaining informed consent for collection of UCB should begin before labor and delivery."

United Synagogue of Conservative Judaism. "Stem Cell Research and Education (2003)," Available online. URL: http://www.uscj.org/Stem_Cell_Research_a6675.html. Downloaded on June 5, 2005. A position statement adopted by the United Synagogue in which it recommends "the use of human embryonic germ and stem cells for research in all appropriate ways."

Vawter, Dorothy E., et al. "The Use of Human Fetal Tissue: Scientific, Ethical, and Policy Concerns." Minneapolis: Center for Biomedical Ethics, University of Minnesota, 1990. Report of a study conducted by the University of Minnesota's Center for Biomedical Ethics on the moral and ethical issues related to the use of fetal tissue for research, transplantation, and other medical purposes.

WEB DOCUMENTS

AmericanCatholic.org. "Stem-cell Research and the Catholic Church." Available online. URL: http://www.americancatholic.org/News/StemCell/default.asp. Downloaded on June 7, 2005. A web site that outlines the Roman Catholic Church's opposition to stem cell research with lengthy citations from letters, reports, and speeches by the Pope, American bishops, and other spokespersons for the Catholic Church.

Bioethics.net. Available online. URL: http://www.bioethics.net. Downloaded on June 5, 2005. A web site maintained by *The American Journal of Bioethics*, containing articles on a wide variety of bioethical issues (including stem cell research), a jobs and events page, a forum, and a blog.

The Center for Bioethics and Human Dignity. "Stem Cell Research." Available online. URL: http://www.cbhd.org/resources/stemcells. Downloaded on May 3, 2005. A collection of articles on the moral and ethical issues surrounding SCR, including subjects such as "The Stem Cell Debate: Are Parthenogenic Human Embryos a Solution?," "Stem Cells & Our Moral Culture," "The Good News and Bad News on Creating Embryos for Research," and "Stem Cell Research: A Constructive Way Forward."

The Coalition of Americans for Research Ethics. "DoNoHarm." Available online. URL: http://www.stemcellresearch.org/index.html. Downloaded on May 16, 2005. An organization founded in 1999 to work against the use of embryonic stem cells in research because it "violates existing law and policy," is "unethical," and is "scientifically unnecessary." The web site contains news on stem cell research, commentary, and information.

Forum on Alternative and Innovative Therapies for Children with Developmental Delays, Brain Injury and Related Neurometabolic Conditions & Disorders. "Cerebral Palsy: Stem Cell Therapy." Available online. URL: http://www.healing-arts.org/children/cp/cpstemcell.htm#M. Downloaded on May 20, 2005. An introduction to stem cell therapy and its potential applications in the treatment of cerebral palsy and related disorders, with a good selection of links to related articles.

Fumento, Michael. "The Stem Cell Cover-Up." Available online. URL: http://www.insightmag.com/media/paper441/news/2004/05/16/National/The-Stem. Cell.CoverUp-682587.shtml. Downloaded on May

5, 2005. The author reviews current efforts to expand the use of embryonic stem cells in research in the United States and suggests that such efforts are driven by economic and personal motivations of SCR researchers. He raises questions about "the questionable morality of a mass campaign to fool the American public."

Human Biotechnology Governance Forum. Available online. URL: http://www.biotechgov.org. Downloaded on May 14, 2005. An online service that provides an extensive array of resources on issues in biotechnology, including stem cell research. The site provides sections on news, commentary, analysis, and resources. A recent search on the topic of "stem cell research" turned up more than 1,000 items in those four areas.

HumGen. "StemGen." Available online. URL: http://www.humgen.umontreal.ca/int/GB_q.cfm?mod=3. A comprehensive list of articles and links to web sites that discuss legal, political, and ethical aspects of stem cell research in all parts of the world.

Klusendort, Scott. "Fetal Tissue and Embryo Stem Cell Research: The March of Dimes, NIH, and Alleged Moral Neutrality." Stand to Reason. Available online. URL: http://www.str.org/free/bioethics/stemcell.pdf. Downloaded on May 12, 2005. An article prepared for the Stand to Reason web site, an organization devoted to "train[ing] Christians to think more clearly about their faith and to make an even-handed, incisive, yet gracious defense for classical Christianity and classical Christian values in the public square." Klusendorf argues that an embryo is "both scientifically and philosophically" a member of the human family, thus making stem cell research morally unacceptable.

LifeNews.com. Available online. URL: http://www.lifenews.com. Downloaded on May 22, 2005. An independent news agency devoted to reporting news related to pro-life issues, such as abortion, assisted suicide, capital punishment, and stem cell research.

Nisbet, Matthew. "The Controversy Over Stem Cell Research and Medical Cloning: Tracking the Rise and Fall of Science in the Public Eye." Available online. URL: http://www.csicop.org/scienceandmedia/controversy. Posted on April 2, 2004. Part 1 of a preliminary report of an on-going study on the forces that shape public opinion on controversial issues, such as cloning and stem cell research, based on an analysis of more than 1,000 articles that have appeared in the print and electronic media.

———. "Understanding What the American Public Really Thinks About Stem Cell and Cloning Research." Available online. URL: http://www.csicop.org/scienceandmedia/controversy/public-opinion.html. Posted on May 2, 2004. Part 2 of the study described in the previous citation.

Ontario Consultants on Religious Tolerance. "Stem Cell Research: All Sides to the Dispute." Available online. URL: http://www.religioustolerance.org/res_stem.htm. Downloaded on May 23, 2005. A web site that attempts to present arguments both for and against stem cell research, with a number of essays and links to other web sites for both sides of the argument.

PollingReport.com. "Science and Nature." Available online. URL: http://www.pollingreport.com/science.htm. Downloaded on May 4, 2005. A web site that summarizes results of about a dozen public opinion polls on the subject of stem cell research. The consensus of the polls seems to be that a substantial majority of Americans currently support the use of federal or state funding to conduct research using embryonic stem cells.

Reagan, Michael. "I'm With My Dad on Stem Cell Research." Human Events.com. Available online. URL: http://www.humaneventsonline.com/article.php?id=4286. Downloaded on May 10, 2005. A 2004 article suggesting that "junk science" has been used to promote the potential of stem cell research and that both he, the author, and his father, former President Ronald Reagan, are strongly opposed to such research.

Smith, Wesley J. "Spinning Stem Cells: A Damning Reporting Pattern." Available online. URL: http://www.nationalreview.com/comment/comment-smith042302.asp. Posted April 23, 2002. The author argues that the press tends to downplay the potential value of adult stem cell research, thereby overestimating the importance of embryonic stem cell research, which he opposes.

"Stem Cell Debate." Justice Talking. Available online. URL: http://www.justicetalking.org/viewprogram.asp?progID=493. Downloaded on May 26, 2005. An online audio recording of a program on issues related to stem cell research in the Justice Talking series of radio programs. The site also contains a listener's guide and suggestions for further reading.

Townhall.com. Available online. URL: http://www.townhall.com. Downloaded on May 11, 2005. A web site that claims to offer "conservative news and information" that contains a number of articles on stem cell research. Enter "stem cell" or "stem cell research" into the site's search function and obtain dozens of articles on SCR and related subjects.

United States Conference of Catholic Bishops. "Stem Cell Research." Available online. URL: http://www.usccb.org/prolife/issues/bioethic/stemcell. Downloaded on May 23, 2005. A web site maintained by the organization's Pro-Life division providing fact sheets on various aspects of stem cell research, copies of letters and testimonials to Congress on the subject, relevant articles in Catholic publications, and news releases on SCR.

Vick, Hannah M. "Embryonic Stem Cell Research: Ethically Wrong Treatment of the Tiniest of Humans." Available online. URL: http://www.cwfa.org/articles/1423/CWA/life/index.htm. Downloaded on June 1, 2005. A discussion of the moral issues that have developed as a result of the rise of embryonic stem cell research, the reasons this author opposes such research, and some satisfactory alternatives to embryonic stem cell research currently available to researchers.

CHAPTER 8

ORGANIZATIONS AND AGENCIES

This chapter contains information on agencies, associations, organizations, and other groups whose primary or exclusive focus involves some aspect of stem cell research. The list of organizations is divided into three general categories: (1) international, federal, state, local, and academic agencies engaged in research on stem cells, in particular, and/or medical research, in general, and in providing information and education about stem cells; (2) organizations interested in promoting the use of stem cell research and/or providing information about this field of scientific research; and (3) groups promoting limitations on stem cell research and/or who are opposed to the development of this field of research.

Stem cell research is a rapidly growing field with a number of academic and research institutions and corporations throughout the world at work on studies of the basic scientific properties of stem cells, their possible medical applications, and their commercial uses. A recent book (*Opportunities in Stem Cell Technology*, Mindbranch, 2002) lists 82 companies involved in the development of stem cell therapeutics, 57 academic organizations involved in stem cell research in the United States, 19 laboratories outside the United States with advanced research in the field, 32 suppliers to the stem cell industry of materials specifically intended for stem cell research, and 10 organizations on the National Institutes of Health list as suppliers having embryonic stem cell lines. This chapter lists only a sample of these organizations, with emphasis on some of the largest, oldest, and most active. For a more complete list of organizations supporting stem cell research, see the web page for Congresswoman Carolyn B. Maloney at http://www.house.gov/maloney/issues/older/stemcellgroups.html.

INTERNATIONAL, FEDERAL, STATE, REGIONAL, AND ACADEMIC ORGANIZATIONS

California Institute for Regenerative Medicine (CIRM)
URL: http://www.cirm.ca.gov
E-mail: info@cirm.ca.gov
Phone: (510) 450-2418
P.O. Box 99740
Emeryville, CA 94662-9740
The California Institute for Regenerative Medicine (CIRM) was created in early 2005 as a consequence of the passage of Proposition 71, the California Stem Cell Research and Cures Initiative in November 2004. That measure provided $3 billion in funding for stem cell research at California universities and research institutions and called for the establishment of a new state agency to make grants and provide loans for stem cell research, research facilities, and other vital research opportunities.

Cambridge Stem Cell Institute
URL: http://www.stemcells.cam.ac.uk
E-mail: contacts@stemcells.cam.ac.uk
Phone: 01223 763366
University of Cambridge
Cambridge CB2 2XY
United Kingdom
The institute brings together researchers with background and interest in stem cell research to improve understanding of the basic science of stem cells with a view to finding ways through which they can be used to solve a variety of medical problems.

The Center for Bioethics at the University of Minnesota
URL: http://www.bioethics.umn.edu
E-mail: bioethx@umn.edu
Phone: (612) 624-9440
N504 Boynton
410 Church Street, SE
Minneapolis, MN 55455
The center's mission is "to advance and disseminate knowledge concerning ethical issues in health care and the life sciences." In order to achieve that mission, the center sponsors original research and scholarship on ethical issues arising out of biological developments; provides courses and other educational opportunities on various topics in bioethics; advises policy makers on issues related to bioethical issues; and publishes documents, holds conferences, answers questions, and provides information to the general public through a variety of outlets.

Center for Regenerative Biology
URL: http://web.uconn.edu/crb/
E-mail: none given; form provided on web page

Phone: (860) 486-6023
University of Connecticut
1392 Storrs Rd.
Unit 4243
Advanced Technology Laboratory Building
Room 116
Storrs, CT 06269-4243
The Center for Regenerative Biology is an interdisciplinary endeavor whose purpose it is to conduct basic research that may lead to the production of new cells, tissues, and organs that can be used as replacement to those damaged by diseases such as diabetes, Parkinson's disease, multiple sclerosis, muscular dystrophy, and many types of cancer.

Center for Stem Cell and Regenerative Medicine
URL: http://ora.ra.cwru.edu/stemcellcenter
E-mail: stemcellcenter@case.edu
Phone: (216) 368-3614
Case Western Reserve University
10900 Euclid Avenue
Cleveland, OH 44106-7284
The Center for Stem Cell Research was established in 2003 with an $19.5 million grant from the state of Ohio. It is composed of researchers from Case Western Reserve University, University Hospitals of Cleveland, The Cleveland Clinic Foundation, Athersys, Inc., and Ohio State University. The center's mission is to utilize adult human stem cells and tissue engineering technology to treat human disease.

Coalition for the Advancement of Medical Research (CAMR)
URL: http://www.camradvocacy.org/fastaction
E-mail: CAMResearch@yahoo.com
Phone: (202) 293-2856
2021 K Street, NW
Suite 305
Washington, DC 20006
The Coalition for the Advancement of Medical Research (CAMR) is an alliance of more than 80 universities, scientific societies, patient-advocacy groups, and individuals interested in life-threatening or debilitating medical conditions, such as Alzheimer's disease, diabetes, brain injury, glaucoma, Parkinson's disease, and Tourette's syndrome. The organization promotes legislative action that will increase research in a variety of fields, including stem cell research, on methods for the treatment and cure of these conditions.

Institute for Stem Cell and Tissue Biology
URL: http://stemcellfacts.ucsf.edu/index.html
E-mail: stemcells@pubaff.ucsf.edu
Phone: See web page
UCSF Medical School, Dean's Office
513 Parnassus Avenue
San Francisco, CA 94143-0410
The Institute for Stem Cell and Tissue Biology was created on September 12, 2005. It combines two existing programs at the University of California at San Francisco Med-

ical School, the program in Developmental and Stem Cell Biology and the UCSF Program in Craniofacial and Mesenchymal Biology. Researchers at the institute work in 60 laboratories attempting to learn more about the nature of stem cells and related cells in humans, mice, zebra fish, and worms, exploring their role in nearly every organ of the body, including the brain, heart, pancreas, liver, blood, bone marrow, skin, prostate, ovaries, and testes.

EuroStemCell
URL: http://www.eurostemcell.org/index.htm
Phone: +44 131 651 7162
Roger Land Building
The King's Buildings
West Mains Road
Edinburgh, Scotland EH9 3JQ
United Kingdom
A consortium of 11 academic institutions and three business enterprises sharing expertise in stem cell technology, developmental biology, tissue repair, cell transplantation, and related fields to discover the basic properties of stem cells and to evaluate their therapeutic potential.

Harvard Stem Cell Institute (HSCI)
URL: http://stemcell.harvard.edu
E-mail: stemcell@harvard.edu
Phone: (617) 496-4050
Harvard University
142 Church Street
Cambridge, MA 02138
HSCI was established in 2004 for the purpose of promoting research on both embryonic and adult stem cells, with particular emphasis on areas with potential for saving human lives and improving human health.

Indiana University Center for Bioethics (IUCB)
URL: http://bioethics.iupui.edu/
E-mail: cntbioet@iupui.edu
Phone: (317) 278-4034
714 North Senate Avenne
Suite EF 200
Indianapolis, IN 46202-3297
The Indiana University Center for Bioethics was established in 2001 for the purpose of promoting the academic and public understanding of bioethics; contributing to the development of social and public policy in health, research, and related fields; and providing support for the furnishing of ethics services at Indiana University hospitals. In October 2002, the center produced a statement on stem cell research, *Diverse Perspectives: Considerations About Embryonic Stem Cell Research*, which is available online at the center's web site at http://bioethics.iupui.edu/Diverse_Perspectives.pdf.

Institute for Stem Cell Research
URL: http://www.iscr.ed.ac.uk
Phone: +44 (0)131 650 5828
The Roger Land Building
Kings Buildings
University of Edinburgh
West Mains Road
Edinburgh, Scotland EH9 3JQ
United Kingdom

A joint program of the Medical Research Council and the University of Edinburgh to promote research on mammalian stem biology, with the goal of developing regenerative therapies that can be used to treat human disease and injury.

Institute of Cancer/Stem Cell Biology and Medicine
URL: http://mednews.stanford.edu/stemcell-index.html
E-mail: See http://mednews.stanford.edu/media_contacts.html for specific addresses
Phone: (650) 723-4000
Stanford University School of Medicine
300 Pasteur Drive
Stanford, CA 94305
Created with a $12 million grant from an anonymous donor in 2002, the institute is touted as "a multidisciplinary initiative that is believed to be the first of its kind in the country." Researchers at the institute have backgrounds in basic and clinical sciences, and use developments in the fields of stem cell biology and cancer biology to advance novel treatments for cancer, diabetes, Parkinson's disease, and cardiovascular disease.

National Human Genome Research Institute (NHGRI)
URL: http://www.genome.gov
Phone: (301) 402-0911
Communications and Public Liaison Branch
National Institutes of Health
Building 31
Room 4B09
31 Center Drive, MSC 2152
9000 Rockville Pike
Bethesda, MD 20892-2152
The National Human Genome Research Institute was originally established in 1989 as the National Center for Human Genome Research (NCHGR) as part of the International Human Genome Project. Its function was to consider moral and ethical issues raised by that project. In 1997, NCHGR was renamed as the National Human Genome Research Institute and elevated to equal status with 26 other institutes within the National Institutes of Health. Most of the center's activities related to stem cell research can be accessed on its Policy and Ethics page (http://www.genome.gov/PolicyEthics), which includes a policy and legislation database; information on privacy and discrimination in genetic-related issues; genetics and the law; health issues; social, cultural and religious issues in genetic research; biomedical research issues; and ethics research.

National Reference Center for Bioethics Literature (NRCBL)
URL: http://www.georgetown.edu/research/nrcbl/nrc/index.htm
E-mail: bioethics@georgetown.edu
Phone: (202) 687-6770
Kennedy Institute of Ethics
Georgetown University
37th & "O" Streets, NW

102 Healy Hall
P.O. Box 571212
Washington, DC 20057-1212
The National Reference Center for Bioethics Literature (NRCBL) grew out of the Ethics Library of the Kennedy Institute of Ethics, which itself was founded in 1973 by a grant from Joseph P. Kennedy, Jr. NRCBL is a collection of more than 28,000 books, 700 periodicals and newsletters, 215,000 journal and newspaper articles, and additional legal materials, regulations, codes, government publications, and other relevant documents concerned with issues in biomedical and professional ethics. The collection is the world's largest related to ethical issues in medicine and biomedical research.

Pittsburgh Development Center
URL: http://www.pdc.magee.edu
E-mail: pdc@pdc.magee.edu
Phone: (412) 641-2400
Magee-Womens Research Institute
204 Craft Avenue
Pittsburgh, PA 15213
The Pittsburgh Development Center of the Magee-Womens Research Institute is a division of the University of Pittsburgh conducting research on the molecular biology of cell function in gametes, embryos, stem cells, assisted reproduction technologies, the origins of developmental diseases, the causes and prevention of adverse pregnancy outcomes, and the potential of stem cells for treating human disease.

President's Council on Bioethics
URL: http://bioethics.gov
E-mail: info@bioethics.gov
Phone: (202) 496-4669
1801 Pennsylvania Avenue, NW
Suite 700
Washington, DC 20006
The President's Council on Bioethics was created by President George W. Bush in 2001 for the purpose of advising him on issues that may emerge as a consequence of advances in biomedical science and technology. The Council has prepared five reports on bioethical issues: *Human Cloning and Human Dignity: An Ethical Inquiry* (July 2002), *Beyond Therapy: Biotechnology and the Pursuit of Happiness* (October 2003), *Being Human: Readings from the President's Council on Bioethics* (December 2003), *Monitoring Stem Cell Research* (January 2004), and *Reproduction and Responsibility: The Regulation of New Biotechnologies* (March 2004).

Stem Cell Institute of New Jersey
URL: http://www.state.nj.us/scitech/stem_home.html
E-mail: njcst@scitech.state.nj.us
Phone: (609) 984-1671
New Jersey Commission on Science and Technology
P. O. Box 832
Trenton, NJ 08625-0832
New Jersey was the second state, after California, to legalize stem cell research in the state. It has authorized an investment of $380 million to create a new research institute at New Brunswick and to finance

research at the institute. The institute will be operated under a joint contract between the University of Medicine and Dentistry of New Jersey and Rutgers University.

Stem Cell Institute of the University of Minnesota Medical School
URL: http://www.stemcell.umn.edu
Phone: (612) 626-4916
MMC 716
420 Delaware Street SE
Minneapolis, MN 55455
The Stem Cell Institute at the University of Minnesota was established in 1999 as the world's first interdisciplinary institute dedicated specifically and exclusively to stem cell research. The institute currently includes 17 schools and centers at the university and focuses on seven diseases primarily: cancer, disorders of the nervous system, cardiology, liver, diabetes, vascular diseases, and genetic disorders. More than 25 researchers work at the institute, which has obtained 15 U.S. patents for its inventions and discoveries.

Stem Cell Network
URL: http://www.stemcellnetwork.ca
Phone: (613) 562-5826
451 Smyth Road
Room 3105
Ottawa, Ontario K1H 8M5
Canada
Stem Cell Network is one of Canada's 22 Networks of Centres of Excellence, administered and funded by the Natural Sciences and Engineering Research Council, the Canadian Institutes of Health Research, and the Social Sciences and Humanities Research Council, in partnership with Industry Canada. The network currently consists of 79 scientists, clinicians, engineers, and ethicists, with the mandate of exploring the therapeutic potential of stem cells for the treatment of diseases currently incurable by conventional approaches. The network intends to invest in multidisciplinary SCR projects, core facilities and relevant technologies; provide matching funds for training of students in the stem cell field; plan ethically responsible approaches for basic and clinical research; create links between the basic science and medical communities; and promote partnerships with industry, health advocacy groups, and other not-for-profit organizations.

Stem Cell Research Center (SCRC)
URL: http://rwjms.umdnj.edu/neuroscience/stem_cell.html
E-mail: spartamb@umdnj.edu
Phone: (732) 235-4029, (732) 235-4990
Department of Neuroscience and Cell Biology
UMDNJ-Robert Wood Johnson Medical School
675 Hoes Lane
Piscataway, NJ 08854
The Stem Cell Research Center (SCRC) was founded in 2002 to study adult stem cells and their potential application in the medical sciences. Researchers at SCRC have

chosen to work with the marrow stromal cell found in the bone marrow of adult animals because it has been shown to have the ability to differentiate into bone, cartilage, tendon, muscle, fat, and nerve cells.

Wisconsin Stem Cell Research Program (WSCRP)
URL: http://www.stemcells.wisc.edu
University of Wisconsin–Madison
1122 Biotechnology Center
425 Henry Mall
Madison, WI 53706-1580

Established after Wisconsin scientist James Thomson first cultured human embryonic stem cells in 1998, WSCRP has four major objectives: (1) fostering campus-wide communication and collaboration among stem cell researchers, (2) providing information on breakthroughs, new publications, seminars, meetings, and funding opportunities in stem-cell related research, (3) establishing a training program for postdoctoral researchers and graduate students, and (4) coordinating fund-raising efforts on stem cell research.

ORGANIZATIONS PROMOTING THE USE OF AND/OR PROVIDING INFORMATION ABOUT STEM CELL RESEARCH

Alliance for Aging Research
URL: http://www.agingresearch.org/
E-mail: info@agingresearch.org
Phone: (202) 293-2856
2021 K Street, NW
Suite 305
Washington, DC 20006

The Alliance for Aging Research was founded in 1986 to promote medical and behavioral research into the aging process, with the goal of helping people live longer and more productive lives. The alliance has taken a strong position in support of all forms of stem cell research, in the expectation that such research may lead to cures for a number of diseases that cause deaths, allowing people to live longer lives. Information on the alliance's stand on stem cell research can be found at http://www.agingresearch.org/healthtopics.cfm#Stem%20Cell%20Research.

Alzheimer's Association
URL: http://www.alz.org/overview.asp
E-mail: info@alz.org
Phone: (800) 272-3900, (312) 335-8700
225 North Michigan Avenue
17th floor
Chicago, IL 60601-7633

The Alzheimer's Association was founded in 1980 with the goal of eliminating Alzheimer's disease through the advancement of scientific research and enhancing the care and support available for individuals with the disease, their families, and their caregivers. The association has taken a stand in support of research on stem cells, outlined and explained on its web page at http://www.alz.org/Research/Papers/stemcell.pdf.

American Cancer Society (ACS)
URL: http://www.cancer.org/docroot/home/index.asp
Phone: (800) 227-2345
1599 Clifton Road, NE
Atlanta, GA 30329
The American Cancer Society (ACS) was founded in New York City by 15 physicians and business leaders in 1913 to find ways of eliminating cancer as a major health problem. ACS is committed to the goal of preventing cancer, saving lives, and reducing the suffering caused by cancer, through research, education, advocacy, and service. The society has taken a position in support of stem cell research and the federal funding thereof. Its position statement is at http://www.cancer.org/docroot/MED/content/MED_2_1x_American_Cancer_Society_Statement_in_Response_to_Presidential_Decision_Regarding_Human_Embryonic_Stem_Cell_Research.asp.

American Heart Association
URL: http://www.american-heart.org/presenter.jhtml?identifier=1200000
Phone: (800) 242-8721
7272 Greenville Avenue
Dallas, TX 75231
The American Heart Association, and its sister organization, the American Stroke Association, work to reduce the number of deaths and disabilities caused by cardiovascular diseases and stroke. Its current policy is to support and fund research conducted with adult stem cells, but not with embryonic stem cells. For more information on the organization's policies and activities in this area, see their web page at http://www.americanheart.org/presenter.jhtml?identifier=4757.

American Medical Association (AMA)
URL: http://www.ama-assn.org
Phone: (800) 621-8335
515 North State Street
Chicago, IL 60610
The American Medical Association (AMA) is the largest professional organization of physicians in the United States. AMA speaks for the medical profession and seeks to advance the quality of medical care in this country. Their supportive position on stem cell research has been expressed in Report 15 of the Council of Scientific Affairs, available online at http://www.ama-assn.org/ama/pub/category/13594.html.

American Society for Cell Biology (ASCB)
URL: http://www.ascb.org/
E-mail: ascbinfo@ascb.org
Phone: (301) 347-9300

8120 Woodmont Avenue
Suite 750
Bethesda, MD 20814-2762
The American Society for Cell Biology (ASCB) was founded in 1960 to promote research in the field of cell biology. The society now claims to have more than 11,000 members in the United States and 50 other nations. The society has taken stands in support of embryonic stem cell research that can be read on the organization's web site. See, for example, its response to President George W. Bush's speech of August 9, 2001 at http://www.ascb.org/newsroom/positionpaper.html.

American Society for Reproductive Medicine (ASRM)
URL: http://www.asrm.org
E-mail: asrm@asrm.org
Phone: (205) 978-5000
1209 Montgomery Highway
Birmingham, AL 35216-2809
Formerly the American Fertility Society, the American Society for Reproductive Medicine (ASRM) is a nonprofit organization founded in 1944 consisting of obstetrician/gynecologists, urologists, reproductive endocrinologists, embryologists, mental health professionals, internists, nurses, practice administrators, laboratory technicians, pediatricians, research scientists, and veterinarians. The goal of the society is to support and sponsor educational activities for the lay public and to provide continuing medical education activities for professionals who are engaged in the practice of and research in reproductive medicine. The society has been active in providing information about and lobbying for cloning and stem cell research. A description of its activities is available on its web page located at http://www.asrm.org/Patients/topics/cloning.html.

American Society of Hematology (ASH)
URL: http://www.hematology.org
E-mail: ash@hematology.org
Phone: (202) 776-0544
1900 M Street, NW
Suite 200
Washington, DC 20036
The first meeting of the American Society of Hematology (ASH) was held in April 1958, with more than 300 specialists in blood-related medical problems in attendance. The organization's mission is "to further the understanding, diagnosis, treatment, and prevention of disorders affecting the blood, bone marrow, and the immunologic, hemostatic and vascular systems, by promoting research, clinical care, education, training, and advocacy in hematology." ASH has taken a strong stand in support of all kinds of stem cell research. Its position paper on this topic can be accessed at http://www.hematology.org/government/policy/avenues_of_stem_cell.cfm.

Bedford Research Foundation
URL: http://www.bedfordresearch.org

E-mail: info@bedfordresearch.org
Phone: (617) 623-5670
P.O. Box 1028
Bedford, MA 01730
The Bedford Research Foundation was founded 1996 by a group of men and women who had contracted HIV/AIDS through blood transfusions with tainted blood. These individuals sought methods for having children through artificial means so as not to pass on their medical condition to the next generation. The Foundation's Special Program of Assisted Reproduction was a result of that effort. In 2002, the foundation decided to become active in the field of stem cell research also and began to seek women to donate eggs for stem cell research projects that had promise to cure diseases that are currently not treatable by any other means.

Biotechnology Industry Organization (BIO)
URL: http://www.bio.org
E-mail: info@bio.org
Phone: (202) 962–9200
1225 Eye Street, NW
Suite 400
Washington, DC 20005
The Biotechnology Industry Organization (BIO) was formed in 1993 by the merger of two existing biotechnology associations, the Industrial Biotechnology Association and the Association of Biotechnology Companies (ABC). The association's objectives are to lobby the industry's positions to elected officials and regulators; to inform national and international media about the industry's progress, its contributions to the quality of life, and its goals and positions; and to provide business development services to member companies, such as information about investments in biotechnology and professional meetings.

The Center for Bioethics and Culture (CBC)
URL: http://www.cbc-network.org
Phone: (510) 594–9000
P.O. Box 20760
Oakland, CA 94620
The Center for Bioethics and Culture is an organization of doctors, nurses, ethicists, clergy, educators, and other professionals interested in promoting an understanding of the ethical and religious implications of biotechnological changes now occurring in the world. The center sponsors seminars and conferences, supports research, and publishes reports and books on issues in the fields of medicine, cloning and stem cell research, reproductive technology, genetics and eugenics, end of life issues, theology and human nature, and biotechnology and public policy.

The Center for Bioethics and Human Dignity (CBHD)
URL: http://www.cbhd.org/aboutcbhd
E-mail: info@cbhd.org
Phone: (847) 317–8180

2065 Half Day Road
Bannockburn, IL 60015
The Center for Bioethics and Human Dignity (CBHD) was founded in 1994 by a group of Christian bioethicists interested in a variety of ethical issues related to biology, including managed care, end-of-life treatment, genetic intervention, euthanasia and suicide, and reproductive technologies. CBHD sponsors an annual international conference on some topic in bioethics and provides speakers for and articles about topics such as biotechnology, cloning, death and dying, genetics, health care and clinical ethics, health and spirituality, alternative medicine, reproductive ethics, and stem cell research.

Center for Genetics and Society (CGS)
URL: http://www.genetics-and-society.org
Phone: (510) 625-0819
436 14th Street
Suite 1302
Oakland, CA 94612
The Center for Genetics and Society (CGS) is a nonprofit organization that works to promote the responsible use of new reproductive technologies that have been and are being developed. Its goal is to work in support of the equitable provision and distribution of health technologies in the United States and throughout the world; for women's health and reproductive rights; for the protection of children; for the rights of the disabled; and for thoughtful and careful application of new reproductive technologies. The CGS web site provides some useful scientific and historical background to the development of stem cell research.

Center for the Study of Technology and Society (TECSOC)
URL: http://www.tecsoc.org
E-mail: washington@tecsoc.org
Phone: (877) 609–5024
1451 Juliana Place
Alexandria, VA 22304
The Center for the Study of Technology and Society (TECSOC) is a tax-exempt organization whose purpose it is to conduct original research and analysis on the way in which technological changes affects society overall. Some areas of special interest to the organization are biotechnology (including stem cell research), national security, and personal security. The organization's web page that includes stem cell information is http://www.tecsoc.org/biotech/biotech.htm.

Christopher Reeve Paralysis Foundation (CRPF)
URL: http://www.christopherreeve.org
Phone: (800) 225-0292
500 Morris Avenue
Springfield, NJ 07081
The Christopher Reeve Paralysis Foundation is on outgrowth of the American Paralysis Association, founded in 1982. CRPF's goal is to provide funding for research aimed to develop treatments and cures for

individuals with spinal cord injury and other central nervous system disorders. As of 1982, the foundation has awarded more than $53 million to fund such research and had made an additional 700 Quality of Life Grants to help improve the daily lives of people living with paralysis, particularly those with spinal cord injuries. The foundation supports and encourages stem cell research as one possible avenue to achieving its goals.

Committee for the Advancement of Stem Cell Research (CASCR)
URL: http://www.cascr.org
Phone: (516) 294–8607
300 Garden City Plaza
Suite 234
Garden City, NY 11530
The Committee for the Advancement of Stem Cell Research was a Section 527 political committee organized to promote the federal funding and promotion of stem cell research in the United States. Its web site provides information on stem cell research and endorsements of candidates who support its views.

Cures Now
URL: http://www.curesnow.org
E-mail: act@CuresNow.org
10100 Santa Monica Boulvard
Suite 1300
Los Angeles, CA 90067
Cures Now promotes the advancement of scientific research in the field of regenerative medicine looking toward the discovery of cures for diseases that are now intractable to all forms of treatment. The organization consists of Nobel laureates; former presidents Jimmy Carter, Gerald Ford, and Bill Clinton; former first lady Nancy Reagan; 125 U.S. medical schools; and friends and relatives of those afflicted with intractable diseases.

Dialogue on Science, Ethics, and Religion (DoSER)
URL: http://www.aaas.org/spp/dser/about/index.shtml
E-mail: doser@aaas.org
Phone: doser@aas.org
American Association for the Advancement of Science
1200 New York Avenue, NW
Washington, DC 20005
The Dialogue on Science, Ethics, and Religion (DoSER) is a division of the American Association for the Advancement of Science established in 1995 to facilitate interaction between the scientific and religious communities. The organization has sponsored national conferences and forums on issues of interest to religious and scientific leaders. It has also developed reports on a number of issues, such as behavioral genetics; consumption, population, and sustainability; the promises and perils of genetic modifications; scientific, historical, philosophical, and theological perspectives on evolution; human inheritable genetic modifications; genetic patenting; and stem cell research and applications.

The Episcopal Church
URL: http://www.episcopalchurch.org/index_new.htm
E-mail: info@episcopalchurch.org
Phone: (800) 334-7626, (212) 716-6000
Episcopal Church Center
815 Second Avenue
New York, NY 10017
The Episcopal Church maintains a web site divided into three major sections: "Life & Work of the Church," "Seekers & Visitors," and "Leadership Resources." Documents relating to the church's support of embryonic stem cell research and related topics can be found at http://www.episcopalchurch.org/ens.htm.

Foundation for Stem Cell Research (FSCR)
URL: http://www.curesforcalifornia.com
E-mail: info@curesforcalifornia.com
Phone: (650) 812-1010
550 California Avenue
Suite 300
Palo Alto, CA 94306
The Foundation for Stem Cell Research (FSCR) formerly the California Research and Cures Coalition, is a coalition of two dozen organizations working to promote stem cell research and to educate the general public about the potential of SCR for treating a variety of medical problems. The coalition consists of groups such as the ALS Association, the American Society for Neurotransplantation and Repair, California Hepatitis C Task Force, the California Medical Association, Catholics for Free Choice, Children's Neurobiological Solutions Foundation, City of Hope, Endocrine Metabolic Medical Center, Gray Panthers California, Hadassah, The Women's Zionist Organization of America, the Michael J. Fox Foundation for Parkinson's Research, and the National Association of Hepatitis Task Forces.

Genetics Policy Institute (GPI)
URL: http://www.genpol.org/index2.htm
Phone: (888) 238-1423
11924 Forest Hill Boulevard
Suite 22
Wellington, FL 33414-6258
The Genetics Policy Institute (GPI) was founded in 2003 as a nonprofit organization with the goal of developing a positive legal framework for "cutting-edge" medical cures, such as stem cell research. The organization states its unequivocal opposition to cloning for the purpose of human reproduction, but supports therapeutic cloning for the purpose of scientific research and the advancement of cures for a host of medical conditions.

International Society for Stem Cell Research (ISSCR)
URL: http://www.isscr.org
E-mail: isscr@isscr.org
Phone: (847) 509-1944
60 Revere Drive
Suite 500
Northbrook, IL 60062

ISSCR was established in 2002 as an independent, nonprofit organization for the purpose of the exchange of information on stem cell research. The society holds an annual meeting to provide an opportunity for leaders in the field to interact with each other; publishes a newsletter, *The Pulse*, that provides researchers with up-to-date information on funding, educational opportunities and scientific advances in the field; and sponsors a number of committees on specific topics, such as development, ethics, government affairs and policy, international activities, junior investigators, membership, planning, public education, publications, and scientific education.

Juvenile Diabetes Research Foundation International (JDRF)
URL: http://www.jdrf.org
E-mail: info@jdrf.org
Phone: (800) 533–2873
120 Wall Street
New York, NY 10005-4001
JDRF's mission is to support research on type 1 (juvenile) diabetes with the goal of finding a cure for the disease and its complications. The foundation believes that stem cell research provides a significant avenue for achieving this goal and supports an expanded use of embryonic stem cells in research projects in the United States. Its position paper on stem cell research is available online at http://www.jdrf.org/files/About_JDRF/StemCellPositionPaper092003.pdf.

The Lasker Foundation
URL: http://www.laskerfoundation.org/index_flash.html
E-mail: info@laskerfoundation.org
Phone: (212) 286-0222
110 East 42nd Street
Suite 1300
New York, NY 10017
The Albert and Mary Lasker Foundation was created to "enlarge public awareness, appreciation, and understanding of promising achievements in medical science in order to increase public support for research." Each year, the foundation recognizes certain outstanding achievements in the field of medical research. The foundation provides a discussion of stem cell research on its web site at http://www.laskerfoundation.org/rprimers/stemcell/stemcell.html.

The Phoebe R. Berman Bioethics Institute
URL: http://www.hopkinsmedicine.org/bioethics
E-mail: bioethic@jhsph.edu
Phone: (410) 955-3018
624 North Broadway
Hampton House 352
Baltimore, MD 21205-1996
The Phoebe R. Berman Bioethics Institute was established at Johns Hopkins University in 1995 for the study of bioethical issues related to developments in health policy, medical care, and the biological, behavioral, and social sciences. One of the institute's areas of special interest—the

program in cell engineering, ethics, and public policy (PCEEPP)—is devoted to analysis of the increasing importance of research on and the development of therapeutic applications for cell engineering (which includes stem cell research) and their ethical implications, and implications for public policy.

Religious Action Center of Reform Judaism (RAC)
URL: http://rac.org/
E-mail: rac@urj.org
Phone: (202) 387-2800
Arthur and Sara Jo Kobacker Building
2027 Massachusetts Avenue, NW
Washington, DC 20036

The Religious Action Center of Reform Judaism (RAC) was established in 1961 to act as "a voice of conscience" to the American Jewish community on legislative and social concerns and to serve as an advocate in Congress on a variety of issues ranging from economic policies in Israel and Russia to civil rights, international peace, and religious liberty. The organization has taken strong stands in support of all forms of stem cell research, stands that are explained and supported on its web site at http://rac.org/advocacy/issues/stemcell/.

The Religious Coalition for Reproductive Choice (RCRC)
URL: http://www.rcrc.org
E-mail: info@rcrc.org
Phone: (202) 628-7700
1025 Vermont Avenue, NW
Suite 1130
Washington, DC 20005

The Religious Coalition for Reproductive Choice (RCRC) was formed in 1973 to ensure that the U.S. Supreme Court's decision on abortion would not be modified or overturned. The coalition is an alliance of national organizations from many faiths; affiliates from throughout the country; and three other national groups—Clergy for Choice Network, Spiritual Youth for Reproductive Freedom, and The Black Church Initiative. The coalition's position in support of all forms of stem cell research, including that involving embryonic and fetal tissue, is outlined on its web page at http://www.rcrc.org/faith_choices/issues/healthcare/stem_cell_research.htm.

Stem Cell Action Network (SCAN)
URL: http://www.stemcellaction.org
6833 Springcrest Circle
Cincinnati, OH 45243

Stem Cell Action Network (SCAN) is a nonprofit, tax-deductible organization of patients and patient advocacy working to promote federal and state funding of stem cell research. The organization has chapters in 16 states, Canada, India, Ireland, Israel, and Lebanon. SCAN is an affiliate of the National Heritage Foundation.

Stem Cell Research Foundation (SCRF)
URL: http://www.stemcellresearchfoundation.org
E-mail: PublicEd@StemCellResearchFoundation.org
Phone: (877) 842-3442
22512 Gateway Center Drive
Clarksburg, MD 20871

The Stem Cell Research Foundation (SCRF) is a program of the American Cell Therapy Research Foundation founded in 2000 to support innovative basic and clinical research in stem cell therapies. Thus far, the program has awarded six research grants totaling $1.8 million.

Student Society for Stem Cell Research (SSSCR)
URL: http://www.stemcellrsch.org
E-mail: info@ssscr.org
303 Bannockburn Avenue
Tampa, FL 33617

Student Society for Stem Cell Research (SSSCR) was founded in August 2003 in the expectation that stem cell research "will revolutionize the field of medicine." It intends to educate the general public about the benefits of SCR and to work "to alleviate human suffering and to promote human health." The group consists of researchers, patient advocates, and policy makers.

Texans for Advancement of Medical Research (TAMR)
URL: http://txamr.org
Phone: (512) 482-9258 (President's contact)

Texans for Advancement of Medical Research (TAMR) consists of scientists, physicians, ethicists, and individuals interested in promoting the benefits of medical research, working for the election of state officials with goals similar to that of TAMR, and advocating on behalf of the citizens of Texas, health advocacy groups, scientists, and physicians. The organization's web site contains a page on stem cell research that includes scientific, economic, and political information on the topic.

Parkinson's Disease Organizations

Although Parkinson's disease is one of the medical conditions most frequently mentioned as amenable to treatment by stem cell therapy, few of the many organizations devoted to advocacy on behalf of Parkinson's patients have taken prominent positions in support of any form of stem cell research. One important document signed by a number of Parkinson's organizations was a Joint Statement on Federal Funding for Embryonic Stem Cell Research issued on August 10, 2001, in response to President George W. Bush's speech of the previous day on stem cell research. That statement can be accessed at http://www.pdf.org/news/news.cfm?selectedItem=75&type=1&returnURL=news.cfm%3Fyear%3D2001%26type%3D1%26start%3D11. Some prominent organizations concerned with Parkinson's disease are listed below.

American Parkinson Disease Association, Inc.
URL: http://www.apdaparkinson.org/user/index.asp
E-mail: apda@apdaparkinson.org
Phone: (800) 223-2732
1250 Hylan Boulevard
Suite 4B
Staten Island, NY 10305
The American Parkinson Disease Association was founded in 1961 to "ease the burden and find a cure" for Parkinson's disease. The organization focuses its work on research, patient support, education, and raising public awareness of the disease.

Michael J. Fox Foundation for Parkinson's Research
URL: http://www.michaeljfox.org
Phone: (800) 708–7644
Grand Central Station
P.O. Box 4777
New York, NY 10163
Actor Michael J. Fox established this foundation in May 2000. He had been diagnosed with a form of Parkinson's disease in 1991, a fact that he made known to the general public in 1998. The foundation is committed to supporting research that will find an early cure for Parkinson's disease. As of January 2005, the foundation has funded more than $50 million in direct and partnered grants for more than 185 Parkinson's research projects in 12 countries.

The Parkinson Alliance
URL: http://www.parkinsonalliance.net/home.html
E-mail: admin@parkinsonalliance.org
Phone: (800) 579-8440
P. O. Box 308
Kingston, NJ 08528-0308
The Parkinson Alliance is a national nonprofit organization dedicated to raising funds for the purpose of financing promising research to find the cause and cure for Parkinson's disease.

National Parkinson Foundation, Inc. (NPF)
URL: http://www.parkinson.org/site/pp.asp?c=9dJFJLPwB&b=71117
E-mail: contact@parkinson.org
Phone: (800) 327-4545, (305) 243-6666
1501 Northwest 9th Avenue/Bob Hope Road
Miami, FL 33136-1494
The National Parkinson Foundation claims to be the "largest and oldest national foundation in the United States." It was founded in 1957 by Jeanne C. Levey for the benefit of her husband, who had the disease. The organization now focuses its energies on the support of research, patient care, education, training, and outreach.

Parkinson's Disease Foundation (PDF)
URL: http://www.pdf.org
E-mail: info@pdf.org
Phone: (800) 457-6676, (212) 923-4700
1359 Broadway
Suite 1509
New York, NY 10018

The Parkinson's Disease Foundation was founded in 1957 by New York City philanthropist and businessman William Black as a way of helping a friend with the disease. The organization claims to be the first national organization created specifically to support research on Parkinson's disease and to assist those struggling with the disorder.

ORGANIZATIONS PROMOTING LIMITATIONS ON AND/OR WHO ARE OPPOSED TO STEM CELL RESEARCH

American Family Association (AFA)
URL: http://www.afa.net
Phone: (662) 844-5036
P.O. Drawer 2440
Tupelo, MS 38803

The American Family Association (AFA) was founded in 1977 by Donald Wildmon to "represent and stand for" traditional family values, especially as those values are perceived to be threatened by television and other forms of the mass media. The web site's search engine provides a number of links to articles and web pages on the subject of stem cell research.

American Life League (ALL)
URL: http://www.all.org
Phone: (540) 659-4171
American Life League
P.O. Box 1350
Stafford, VA 22555

Members of the American Life League (ALL) believe that life begins at the moment of fertilization and that all lives are equally sacred. They are opposed to all forms of abortion under all circumstances, including rape, incest, or risk to a pregnant woman. Because of this philosophy, they also believe that all forms of embryonic stem cell research should be prohibited. For further information on the organization's stand on stem cell research, see http://search.atomz.com/search/?sp-q=stem+cell+research&sp-a=sp0611bd00&sp-advanced=1&sp-p=all&sp-w-control=1&sp-d=custom&sp-date-range=-1&sp-start-month=0&sp-start-day=0&sp-start-year=&sp-end-month=0&sp-end-day=0&sp-end-year=&sp-f=iso-8859-1.

Americans to Ban Cloning (ABC)
URL: http://cloninginformation.org
E-mail: media@cloninginformation.org
Phone: (202) 347-6840
1100 H Street, NW
Suite 700
Washington, DC 20005

An organization of unidentified membership whose purpose it is to promote an absolute ban on all types of human cloning in the United States and other nations of the world. The web site provides excellent articles and positions statements that argue in support of such a ban.

Christian Coalition of America (CCA)
URL: http://www.cc.org
E-mail: Coalition@cc.org
Phone: (202) 479-6900
P.O. Box 37030
Washington, DC 20013-7030
The Christian Coalition of America (CCA) was founded by Pat Robertson in 1989 to provide a voice for conservative Christians who wished to make their views sought in the electoral process. The organization supports and promotes individuals whose viewpoints are consistent with its own and claims to have sent out 70 million "voter guides" in the November 2004 elections to help elect such individuals. The organization has taken a stand in opposition to all forms of embryonic stem cell research, a position explained in detail in the "Issues" section of its web site.

Comment on Reproductive Ethics (CORE)
URL: http://www.corethics.org
E-mail: info@corethics.org
Phone: 00 44 (0) 207 5812623
P.O. Box 4593
London SW3 6XE
United Kingdom
Comment on Reproductive Ethics (CORE) was founded in 1994 by 16 individuals in Great Britain to promote study and consideration of "ethical dilemmas surrounding human reproduction, particularly the new technologies of assisted conception." Although the organization claims to encourage "informed and balanced debate" over such issues, its underlying assumption and starting point for such a debate is "[a]bsolute respect for the human embryo."

Concerned Women for America (CWA)
URL: http://www.cwfa.org/main.asp
Phone: (202) 488-7000
1015 Fifteenth Street, NW
Suite 1100
Washington, DC 20005
An organization founded in 1978 by Beverly LaHaye to provide counterbalance to what she saw as anti-God, anti-family feminism. CWA's mission is to "protect and promote Biblical values among all citizens—first through prayer, then education, and finally by influencing our society—thereby reversing the decline in moral values in our nation." Links to articles and information on one of the organization's current "hot topics," stem cell research, is provided at http://www.cwfa.org/hot-topics.asp.

Council for Responsible Genetics (CRG)
URL: http://www.gene-watch.org
E-mail: crg@gene-watch.org

Phone: (617) 868-0870
5 Upland Road
Suite 3
Cambridge, MA 02140
The Council for Responsible Genetics (CRG) is a nonprofit, nongovernmental agency founded in 1983 to promote public consideration of and debate about the social, ethical, and environmental implications of advances in a wide range of fields involving genetic research. The council has chosen to take a stand against embryonic stem cell research and has published position papers, articles, and reports in support of this position. CRG publishes a regular magazine, *Gene Watch*, with information about its programs and about genetic research and projects of interest to the general public.

Culture of Life Foundation & Institute
URL: http://www.christianity.com/cultureoflife
E-mail: clf@culture-of-life.org
Phone: (888) 865-5433, (202) 289-2500
1413 K Street, NW
Suite 1000
Washington, DC 20005
The Culture of Life Foundation & Institute is a Roman Catholic organization interested in analyzing, interpreting, and communicating information about advances in science and technology related to human life, defined as the event that occurs at the moment of conception. The organization publishes a regular newsletter, *Culture & Cosmos*, and an "Adult Stem Cell Breakthrough Bulletin" that carries recent developments in the field of adult stem cell research.

Do No Harm: The Coalition of Americans for Research Ethics
URL: http://www.stemcellresearch.org
Phone: (202) 347-6840
1100 H Street, NW
Suite 700
Washington, DC 20005
Do No Harm's founding statement says that stem cell research "is a worthy scientific priority as long as we pursue it ethically," but opposes embryonic stem cell research because (1) it violates existing law and policy, (2) it is unethical, and (3) its is scientifically unnecssary.

Focus on the Family
URL: http://www.family.org
Phone: (800) 232-6459
Colorado Springs, CO 80995
The mission of Focus on the Family is "to cooperate with the Holy Spirit in disseminating the Gospel of Jesus Christ to as many people as possible, and, specifically, to accomplish that objective by helping to preserve traditional values and the institution of the family." The organization maintains a section on its web site with articles about stem cell research. Use the search function to find articles on "stem cell research."

National Right to Life Committee (NRLC)
URL: http://www.nrlc.org
E-mail: NRLC@nrlc.org

Phone: (202) 626-8800
512 10th Street, NW
Washington, DC 20004
The National Right to Life Committee (NRLC) was formed in 1973 in response to the U.S. Supreme Court's decision on abortion. The organization currently has more than 3,000 chapters in the 50 states and the District of Columbia. Its ultimate goal is to "restore legal protection to innocent human life" by banning all types of abortions under all circumstances. In addition to its interest in abortion, NRLC is concerned with other issues of medical ethics related to abortion, such as euthanasia and infanticide. The organization's position on stem cell research is outlined and supported with related documents on its web page at http://www.nrlc.org/Killing_Embryos/index.html.

Nightlight Christian Adoptions (NCA)
URL: http://www.nightlight.org
E-mail: Info@Nightlight.org
Phone: (714) 278-1020
801 East Chapman Avenue
Suite 106
Fullerton, CA 92831
Nightlight Christian Adoptions (NCA) is a nonprofit organization chartered in the state of California that attempts to "bring glory and honor to our Lord and Savior, Jesus Christ" through providing domestic and international adoption services and counseling services to those who find themselves unexpectedly pregnant. The program's Snowflake Adoption Program received national attention in 2005 when President George W. Bush introduced a number of babies who had been born from frozen embryos no longer wanted by donors and donated to the NCA Snowflake embryo adoption program. The program's goal is to find homes for as many as possible of the estimated 400,000 frozen embryos currently being stored in in vitro fertilization program storage banks.

Secretariat for Pro-Life Activities
United States Conference of Catholic Bishops
URL: http://www.usccb.org/prolife
Phone: (202) 541-3070
3211 4th Street, NE
Washington, DC 20017-1194
The expressed goal of the Secretariat for Pro-Life Activities is to teach respect for all human life, beginning from conception to natural death, and to organize for its protection. The secretariat's web site provides information and resources on a variety of topics, including abortion, assisted suicide, capital punishment, cloning, contraception, embryo research, euthanasia, fetal research, in vitro fertilization, the morning-after pill, partial-birth abortion, post abortion, RU-486, stem cell research, unborn victims of violence, and women.

The Southern Baptist Convention, Ethics and Religious Liberty Commission
URL: http://www.erlc.com

E-mail: See "Contact Us" page at
http://www.faithandfamily.com/CC_Content_Page/0,,PTID314166|CHID602940|CIID,00.html
Phone: (202) 547-8105
505 Second Street, NE
Washington, DC 20002
The Ethics and Religious Liberty Commission is the policy arm of the Southern Baptist Convention, organized to address social, moral, and ethical concerns related to the Baptist faith. One of the issues on which the commission has taken a stand is stem cell research, which it opposes when embryonic stem cells are involved. A web page containing documents on this topic is located at http://sites.silaspartners.com/CC/CDA/CC_Content/CC_Archive_Display_Page/0,,PTID314166CHID599214 GRPID31256,00.html.

Stand to Reason
URL: http://www.str.org
E-mail: questions@str.org
Phone: (562) 595-7333
1438 East 33rd Street
Signal Hill, CA 90755
Stand to Reason is a group of individuals whose goal it is "to provide the training to build a new generation of confident, courageous, yet winsome and attractive ambassadors for Chirst capable of restoring credibilty to the Christian world view." The organization's web site contains a number of articles that argue against the use of stem cells because such research results in the death of what is claimed to be a human life.

PART III

APPENDICES

APPENDIX A

DIAGRAMS

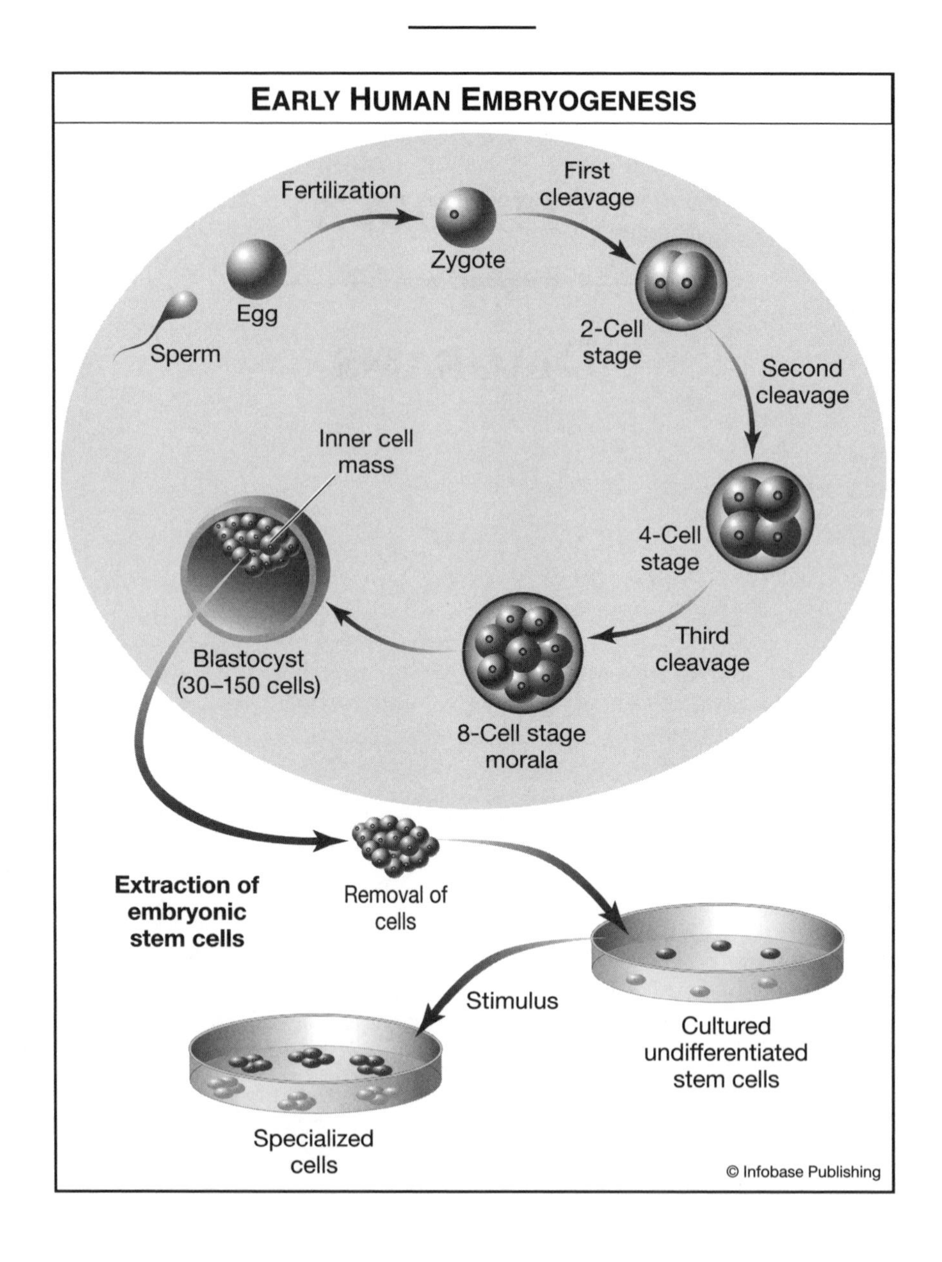
EARLY HUMAN EMBRYOGENESIS
Fertilization
First cleavage
Zygote
Egg
Sperm
2-Cell stage
Second cleavage
Inner cell mass
4-Cell stage
Third cleavage
Blastocyst (30–150 cells)
8-Cell stage morala
Extraction of embryonic stem cells
Removal of cells
Stimulus
Cultured undifferentiated stem cells
Specialized cells
© Infobase Publishing

Appendix A

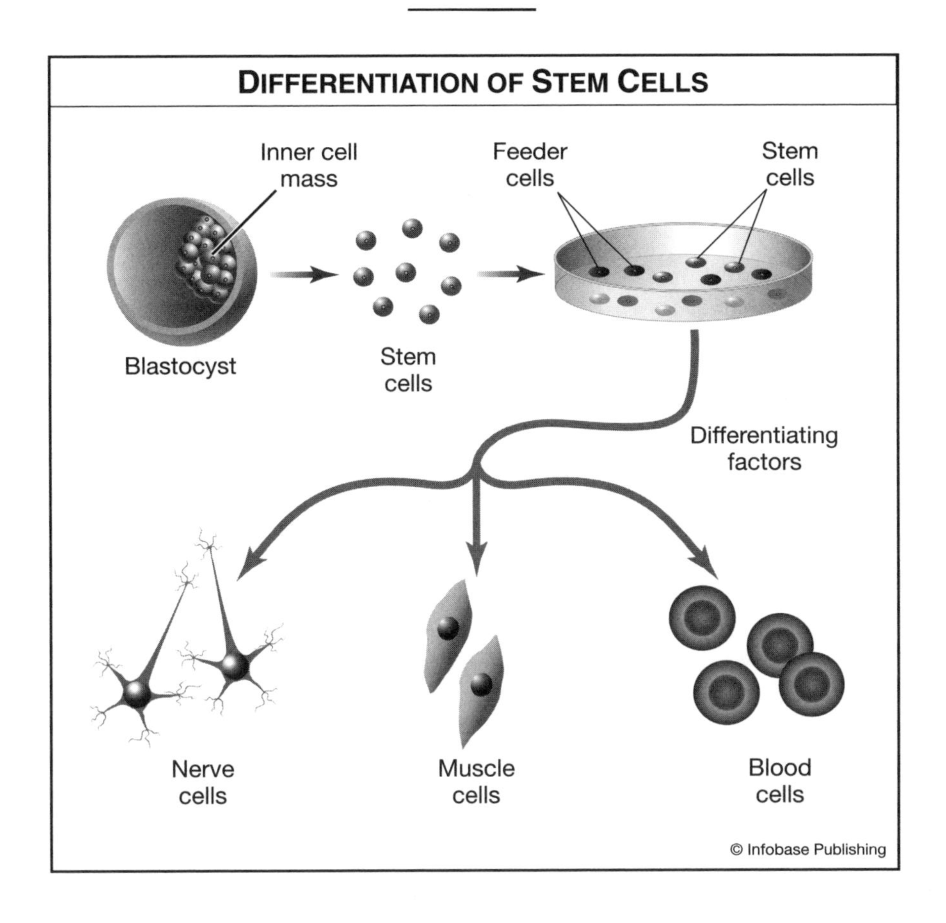
DIFFERENTIATION OF STEM CELLS
Inner cell mass
Feeder cells
Stem cells
Blastocyst
Stem cells
Differentiating factors
Nerve cells
Muscle cells
Blood cells
© Infobase Publishing

APPENDIX B

ROE V. WADE ON THE LEGAL STATUS OF THE EMBRYO, 410 U.S. 113, 1973

Note: The first case in which the U.S. Supreme Court dealt specifically with the legal status of the embryo was Roe v. Wade, *a case in which the primary issue was a woman's right to have an abortion. In deciding that case, the Court first had to determine whether the embryo or fetus that was to be aborted was a person or not. If it were, it had all the legal rights of any citizen and could not be deprived of its life. If it were not, it did not have those legal rights. The Court began by reviewing the status of the embryo and fetus in English common law. (References are omitted from this extract.)*

It is undisputed that at common law, abortion performed before "quickening"—the first recognizable movement of the fetus in utero, appearing usually from the 16th to the 18th week of pregnancy—was not an indictable offense. The absence of a common-law crime for pre-quickening abortion appears to have developed from a confluence of earlier philosophical, theological, and civil and canon law concepts of when life begins. These disciplines variously approached the question in terms of the point at which the embryo or fetus became "formed" or recognizably human, or in terms of when a "person" came into being, that is, infused with a "soul" or "animated." A loose consensus evolved in early English law that these events occurred at some point between conception and live birth. This was "mediate animation." Although Christian theology and the canon law came to fix the point of animation at 40 days for a male and 80 days for a female, a view that persisted until the 19th century, there was otherwise little agreement about the precise time of formation or animation. There was agreement, however, that prior to this point the fetus was to be regarded as part of the mother, and its destruction, therefore, was not homicide. Due to continued uncertainty about the precise time when animation occurred, to the lack of any empirical basis for the 40–80-day view, and perhaps to Aquinas' definition of movement as one of the two first principles of life, Bracton focused upon quickening as the critical point. The significance of quickening was echoed by later common-law scholars and found its way into the received common law in this country.

[The Court then explored references made to "personhood" in the U.S. Constitution.]

The Constitution does not define "person" in so many words. Section 1 of the Fourteenth Amendment contains three references to "person." The first, in defining "citizens," speaks of "persons born or naturalized in the United States." The word also appears both in the Due Process Clause and in the Equal Protection Clause. "Person" is used in other places in the Constitution: in the listing of qualifications for Representatives and Senators, Art. I, 2, cl. 2, and 3, cl. 3; in the Apportionment Clause, Art. I, 2, cl. 3; in the Migration and Importation provision, Art. I, 9, cl. 1; in the Emolument Clause, Art. I, 9, cl. 8; in the Electors provisions, Art. II, 1, cl. 2, and the superseded cl. 3; in the provision outlining qualifications for the office of President, Art. II, 1, cl. 5; in the Extradition provisions, Art. IV, 2, cl. 2, and the superseded Fugitive Slave Clause 3; and in the Fifth, Twelfth, and Twenty-second Amendments, as well as in 2 and 3 of the Fourteenth Amendment. But in nearly all these instances, the use of the word is such that it has application only postnatally. None indicates, with any assurance, that it has any possible pre-natal application.

[Based on this review, the Court then outlined its position on the legal status of the embryo and fetus.]

All this, together with our observation, supra, that throughout the major portion of the 19th century prevailing legal abortion practices were far freer than they are today, persuades us that the word "person," as used in the Fourteenth Amendment, does not include the unborn. This is in accord with the results reached in those few cases where the issue has been squarely presented.

[The Court listed seven such cases.]

Indeed, our decision in *United States v. Vuitch*, 402 U.S. 62 (1971), inferentially is to the same effect, for we there would not have indulged in statutory interpretation favorable to abortion in specified circumstances if the necessary consequence was the termination of life entitled to Fourteenth Amendment protection.

APPENDIX C

DAVIS V. DAVIS ON THE LEGAL STATUS OF THE EMBRYO, SUPREME COURT OF TENNESSEE, 842 S.W.2D 588, 597, 1992

Note: Davis v. Davis *is the first case in which a U.S. court was asked to rule on the legal status of embryos, created by in vitro fertilization, over which the gamete donors disagreed as to their ultimate disposition. The court begins by attempting to determine whether the embryos are persons or property. Citations are omitted from this extract.*

One of the fundamental issues the inquiry poses is whether the preembryos in this case should be considered "persons" or "property" in the contemplation of the law. The Court of Appeals held, correctly, that they cannot be considered "persons" under Tennessee law:

[46] The policy of the state on the subject matter before us may be gleaned from the state's treatment of fetuses in the womb. . . . The state's Wrongful Death Statute, Tenn. Code Ann. § 20-5-106 does not allow a wrongful death for a viable fetus that is not first born alive. Without live birth, the Supreme Court has said, a fetus is not a "person" within the meaning of the statute. Other enactments by the legislature demonstrate even more explicitly that viable fetuses in the womb are not entitled to the same protection as "persons." Tenn. Code Ann. § 39-15-201 incorporates the trimester approach to abortion outlined in *Roe v. Wade*, 410 U.S. 113 (1973). A woman and her doctor may decide on abortion within the first three months of pregnancy but after three months, and before viability,

abortion may occur at a properly regulated facility. Moreover, after viability, abortion may be chosen to save the life of the mother. This statutory scheme indicates that as embryos develop, they are accorded more respect than mere human cells because of their burgeoning potential for life. But, even after viability, they are not given legal status equivalent to that of a person already born. This concept is echoed in Tennessee's murder and assault statutes, which provide that an attack or homicide of a viable fetus may be a crime but abortion is not.

[48] Nor do preembryos enjoy protection as "persons" under federal law. In *Roe v. Wade*, 410 U.S. 113 (1973), the United States Supreme Court explicitly refused to hold that the fetus possesses independent rights under law, based upon a thorough examination of the federal constitution, relevant common law principles, and the lack of scientific consensus as to when life begins. The Supreme Court concluded that "the unborn have never been recognized in the law as persons in the whole sense." As a matter of constitutional law, this Conclusion has never been seriously challenged. Hence, even as the Supreme Court in *Webster v. Reproductive Health Services*, 492 U.S. 490 (1989), permitted the states some additional leeway in regulating the right to abortion established in *Roe v. Wade*, the Webster decision did no more than recognize a compelling state interest in potential life at the point when viability is possible. Thus, as Justice O'Connor noted, "viability remains the 'critical point.'" That stage of fetal developmental is far removed, both qualitatively and quantitatively, from that of the four- to eight-cell preembryos in this case.

[49] Left undisturbed, the trial court's ruling would have afforded preembryos the legal status of "persons" and vested them with legally cognizable interests separate from those of their progenitors. Such a decision would doubtless have had the effect of outlawing IVF programs in the state of Tennessee. But in setting aside the trial court's judgment, the Court of Appeals, at least by implication, may have swung too far in the opposite direction.

[50] The intermediate court, without explicitly holding that the preembryos in this case were "property," nevertheless awarded "joint custody" of them to Mary Sue Davis and Junior Davis, citing T.C.A. §§ 68-30-101 and 39-15-208, and *York v. Jones*, 717 F.Supp. 421 (E.D. Va. 1989), for the proposition that "the parties share an interest in the seven fertilized ova." The intermediate court did not otherwise define this interest.

[51] The provisions of T.C.A. §§ 68-30-101 et seq., on which the intermediate appellate court relied, codify the Uniform Anatomical Gift Act. T.C.A. § 39-15-208 prohibits experimentation or research using an aborted fetus in the absence of the woman's consent. These statutes address the question of who controls Disposition of human organs and tissue with no further potential for autonomous human life; they are not precisely

controlling on the question before us, because the "tissue" involved here does have the potential for developing into independent human life, even if it is not yet legally recognizable as human life itself.

[52] The intermediate court's reliance on *York v. Jones*, is even more troublesome. That case involved a dispute between a married couple undergoing IVF procedures at the Jones Institute for Reproductive Medicine in Virginia. When the Yorks decided to move to California, they asked the Institute to transfer the one remaining "frozen embryo" that they had produced to a fertility clinic in San Diego for later implantation. The Institute refused and the Yorks sued. The federal district court assumed without deciding that the subject matter of the dispute was "property." The York court held that the "cryopreservation agreement" between the Yorks and the Institute created a bailment relationship, obligating the Institute to return the subject of the bailment to the Yorks once the purpose of the bailment had terminated.

[53] In this case, by citing to *York v. Jones* but failing to define precisely the "interest" that Mary Sue Davis and Junior Davis have in the preembryos, the Court of Appeals has left the implication that it is in the nature of a property interest. For purposes of clarity in future cases, we conclude that this point must be further addressed.

[54] To our way of thinking, the most helpful Discussion on this point is found not in the minuscule number of legal opinions that have involved "frozen embryos," but in the ethical standards set by The American Fertility Society, as follows:

[55] Three major ethical positions have been articulated in the debate over preembryo status. At one extreme is the view of the preembryo as a human subject after fertilization, which requires that it be accorded the rights of a person. This position entails an obligation to provide an opportunity for implantation to occur and tends to ban any action before transfer that might harm the preembryo or that is not immediately therapeutic, such as freezing and some preembryo research.

[56] At the opposite extreme is the view that the preembryo has a status no different from any other human tissue. With the consent of those who have decision-making authority over the preembryo, no limits should be imposed on actions taken with preembryos.

[57] A third view—one that is most widely held—takes an intermediate position between the other two. It holds that the preembryo deserves respect greater than that accorded to human tissue but not the respect accorded to actual persons. The preembryo is due greater respect than other human tissue because of its potential to become a person and because of its symbolic meaning for many people. Yet, it should not be treated as a person, because it has not yet developed the features of personhood, is not yet

established as developmentally individual, and may never realize its biologic potential.

[Based on this analysis, the court decides that:]

We conclude that preembryos are not, strictly speaking, either "persons" or "property," but occupy an interim category that entitles them to special respect because of their potential for human life. It follows that any interest that Mary Sue Davis and Junior Davis have in the preembryos in this case is not a true property interest. However, they do have an interest in the nature of ownership, to the extent that they have decision-making authority concerning Disposition of the preembryos, within the scope of policy set by law.

[Finally, in its summary of the case, the court outlines its position on the legal status of embryos created by in vitro fertilization whose ultimate fate is in dispute between gamete donors.]

In summary, we hold that disputes involving the Disposition of preembryos produced by in vitro fertilization should be resolved, first, by looking to the preferences of the progenitors. If their wishes cannot be ascertained, or if there is dispute, then their prior agreement concerning Disposition should be carried out. If no prior agreement exists, then the relative interests of the parties in using or not using the preembryos must be weighed. Ordinarily, the party wishing to avoid procreation should prevail, assuming that the other party has a reasonable possibility of achieving parenthood by means other than use of the preembryos in question. If no other reasonable alternatives exist, then the argument in favor of using the preembryos to achieve pregnancy should be considered. However, if the party seeking control of the preembryos intends merely to donate them to another couple, the objecting party obviously has the greater interest and should prevail.

[113] But the rule does not contemplate the creation of an automatic veto, and in affirming the judgment of the Court of Appeals, we would not wish to be interpreted as so holding.

[114] For the reasons set out above, the judgment of the Court of Appeals is affirmed, in the appellee's favor. This ruling means that the Knoxville Fertility Clinic is free to follow its normal procedure in dealing with unused preembryos, as long as that procedure is not in conflict with this opinion. Costs on appeal will be taxed to the appellant.

APPENDIX D

THE DICKEY AMENDMENT, 1995

Note: This has been adopted each year since 1995, though not in exactly the same form.

Section 510.
(a) None of the funds made available in this Act may be used for —

(1) the creation of a human embryo or embryos for research purposes; or
(2) research in which a human embryo or embryos are destroyed, discarded, or knowingly subjected to risk of injury or death greater than that allowed for research on fetuses in utero under 45 CFR 46.208(a)(2) and section 498(b) of the Public Health Service Act (42 U.S.C. 289g(b)).

(b) For purposes of this section, the term 'human embryo or embryos' includes any organism, not protected as a human subject under 45 CFR 46 as of the date of the enactment of this Act, that is derived by fertilization, parthenogenesis, cloning, or any other means from one or more human gametes or human diploid cells.

Source: "White Paper: Alternative Sources of Pluripotent Stem Cells." The President's Council on Bioethics. Available online. URL: http://www.bioethics.gov/reports/white_paper/text.html#_edn10. Downloaded on July 22, 2005.

APPENDIX E

STATEMENT OF POLICY ON STEM CELL RESEARCH BY PRESIDENT GEORGE W. BUSH, AUGUST 9, 2001

Good evening. I appreciate you giving me a few minutes of your time tonight so I can discuss with you a complex and difficult issue, an issue that is one of the most profound of our time.

The issue of research involving stem cells derived from human embryos is increasingly the subject of a national debate and dinner table discussions. The issue is confronted every day in laboratories as scientists ponder the ethical ramifications of their work. It is agonized over by parents and many couples as they try to have children, or to save children already born.

The issue is debated within the church, with people of different faiths, even many of the same faith coming to different conclusions. Many people are finding that the more they know about stem cell research, the less certain they are about the right ethical and moral conclusions.

My administration must decide whether to allow federal funds, your tax dollars, to be used for scientific research on stem cells derived from human embryos. A large number of these embryos already exist. They are the product of a process called in vitro fertilization, which helps so many couples conceive children. When doctors match sperm and egg to create life outside the womb, they usually produce more embryos than are planted in the mother. Once a couple successfully has children, or if they are unsuccessful, the additional embryos remain frozen in laboratories.

Some will not survive during long storage; others are destroyed. A number have been donated to science and used to create privately funded stem cell lines. And a few have been implanted in an adoptive mother and born, and are today healthy children.

Based on preliminary work that has been privately funded, scientists believe further research using stem cells offers great promise that could help improve the lives of those who suffer from many terrible diseases—from juvenile diabetes to Alzheimer's, from Parkinson's to spinal cord injuries. And while scientists admit they are not yet certain, they believe stem cells derived from embryos have unique potential.

You should also know that stem cells can be derived from sources other than embryos—from adult cells, from umbilical cords that are discarded after babies are born, from human placenta. And many scientists feel research on these type of stem cells is also promising. Many patients suffering from a range of diseases are already being helped with treatments developed from adult stem cells.

However, most scientists, at least today, believe that research on embryonic stem cells offer [*sic*] the most promise because these cells have the potential to develop in all of the tissues in the body.

Scientists further believe that rapid progress in this research will come only with federal funds. Federal dollars help attract the best and brightest scientists. They ensure new discoveries are widely shared at the largest number of research facilities and that the research is directed toward the greatest public good.

The United States has a long and proud record of leading the world toward advances in science and medicine that improve human life. And the United States has a long and proud record of upholding the highest standards of ethics as we expand the limits of science and knowledge. Research on embryonic stem cells raises profound ethical questions, because extracting the stem cell destroys the embryo, and thus destroys its potential for life. Like a snowflake, each of these embryos is unique, with the unique genetic potential of an individual human being.

As I thought through this issue, I kept returning to two fundamental questions: First, are these frozen embryos human life, and therefore, something precious to be protected? And second, if they're going to be destroyed anyway, shouldn't they be used for a greater good, for research that has the potential to save and improve other lives?

I've asked those questions and others of scientists, scholars, bioethicists, religious leaders, doctors, researchers, members of Congress, my Cabinet, and my friends. I have read heartfelt letters from many Americans. I have given this issue a great deal of thought, prayer and considerable reflection. And I have found widespread disagreement.

On the first issue, are these embryos human life—well, one researcher told me he believes this five-day-old cluster of cells is not an embryo, not yet an individual, but a pre-embryo. He argued that it has the potential for life, but it is not a life because it cannot develop on its own.

An ethicist dismissed that as a callous attempt at rationalization. Make no mistake, he told me, that cluster of cells is the same way you and I, and all the rest of us, started our lives. One goes with a heavy heart if we use these, he said, because we are dealing with the seeds of the next generation.

And to the other crucial question, if these are going to be destroyed anyway, why not use them for good purpose—I also found different answers. Many argue these embryos are byproducts of a process that helps create life, and we should allow couples to donate them to science so they can be used for good purpose instead of wasting their potential. Others will argue there's no such thing as excess life, and the fact that a living being is going to die does not justify experimenting on it or exploiting it as a natural resource.

At its core, this issue forces us to confront fundamental questions about the beginnings of life and the ends of science. It lies at a difficult moral intersection, juxtaposing the need to protect life in all its phases with the prospect of saving and improving life in all its stages.

As the discoveries of modern science create tremendous hope, they also lay vast ethical mine fields. As the genius of science extends the horizons of what we can do, we increasingly confront complex questions about what we should do. We have arrived at that brave new world that seemed so distant in 1932, when Aldous Huxley wrote about human beings created in test tubes in what he called a "hatchery."

In recent weeks, we learned that scientists have created human embryos in test tubes solely to experiment on them. This is deeply troubling, and a warning sign that should prompt all of us to think through these issues very carefully.

Embryonic stem cell research is at the leading edge of a series of moral hazards. The initial stem cell researcher was at first reluctant to begin his research, fearing it might be used for human cloning. Scientists have already cloned a sheep. Researchers are telling us the next step could be to clone human beings to create individual designer stem cells, essentially to grow another you, to be available in case you need another heart or lung or liver.

I strongly oppose human cloning, as do most Americans. We recoil at the idea of growing human beings for spare body parts, or creating life for our convenience. And while we must devote enormous energy to conquering disease, it is equally important that we pay attention to the moral concerns raised by the new frontier of human embryo stem cell research. Even the most noble ends do not justify any means.

My position on these issues is shaped by deeply held beliefs. I'm a strong supporter of science and technology, and believe they have the potential for incredible good—to improve lives, to save life, to conquer disease. Research offers hope that millions of our loved ones may be cured of a disease and rid of their suffering. I have friends whose children suffer from juvenile diabetes. Nancy Reagan has written me about President Reagan's struggle with Alzheimer's. My own family has confronted the tragedy of childhood leukemia. And, like all Americans, I have great hope for cures.

I also believe human life is a sacred gift from our Creator. I worry about a culture that devalues life, and believe as your President I have an important obligation to foster and encourage respect for life in America and throughout the world. And while we're all hopeful about the potential of this research, no one can be certain that the science will live up to the hope it has generated.

Eight years ago, scientists believed fetal tissue research offered great hope for cures and treatments—yet, the progress to date has not lived up to its initial expectations. Embryonic stem cell research offers both great promise and great peril. So I have decided we must proceed with great care.

As a result of private research, more than 60 genetically diverse stem cell lines already exist. They were created from embryos that have already been destroyed, and they have the ability to regenerate themselves indefinitely, creating ongoing opportunities for research. I have concluded that we should allow federal funds to be used for research on these existing stem cell lines, where the life and death decision has already been made.

Leading scientists tell me research on these 60 lines has great promise that could lead to breakthrough therapies and cures. This allows us to explore the promise and potential of stem cell research without crossing a fundamental moral line, by providing taxpayer funding that would sanction or encourage further destruction of human embryos that have at least the potential for life.

I also believe that great scientific progress can be made through aggressive federal funding of research on umbilical cord placenta, adult and animal stem cells which do not involve the same moral dilemma. This year, your government will spend $250 million on this important research.

I will also name a President's council to monitor stem cell research, to recommend appropriate guidelines and regulations, and to consider all of the medical and ethical ramifications of biomedical innovation. This council will consist of leading scientists, doctors, ethicists, lawyers, theologians and others, and will be chaired by Dr. Leon Kass, a leading biomedical ethicist from the University of Chicago.

This council will keep us apprised of new developments and give our nation a forum to continue to discuss and evaluate these important issues. As we go forward, I hope we will always be guided by both intellect and heart, by both our capabilities and our conscience.

I have made this decision with great care, and I pray it is the right one. Thank you for listening. Good night, and God bless America.

Source: "Remarks by the President on Stem Cell Research." The White House. Available online. URL: http://www.whitehouse.gov/news/releases/2001/08/20010809-2.html. Downloaded on July 21, 2005.

APPENDIX F

RECOMMENDATIONS: ETHICAL AND POLICY ISSUES IN RESEARCH INVOLVING HUMAN PARTICIPANTS, NATIONAL BIOETHICS ADVISORY COMMISSION, AUGUST 2001

SUMMARY OF RECOMMENDATIONS

Note: Recommendations are numbered according to the chapter from which they are taken in the overall report. Since there are no recommendations in Chapter 1, there are no recommendations with the designation "*Recommendation 1.x.*"

Recommendation 2.1: The federal oversight system should protect the rights and welfare of human research participants by requiring 1) independent review of risks and potential benefits and 2) voluntary informed consent. Protection should be available to participants in both publicly and privately sponsored research. Federal legislation should be enacted to provide such protection.

Recommendation 2.2: To ensure the protection of the rights and welfare of all research participants, federal legislation should be enacted to create a single, independent federal office, the National Office for Human Research Oversight (NOHRO), to lead and coordinate the oversight system. This office should be responsible for policy development, regulatory reform (see Recommendation 2.3), research review and monitoring, research ethics education, and enforcement.

Appendix F

Recommendation 2.3: A unified, comprehensive federal policy embodied in a single set of regulations and guidance should be created that would apply to all types of research involving human participants (see Recommendation 2.2).

Recommendation 2.4: Federal policy should cover research involving human participants that entails systematic collection or analysis of data with the intent to generate new knowledge. Research should be considered to involve human participants when individuals 1) are exposed to manipulations, interventions, observations, or other types of interactions with investigators or 2) are identifiable through research using biological materials, medical and other records, or databases. Federal policy also should identify those research activities that are not subject to federal oversight and outline a procedure for determining whether a particular study is or is not covered by the oversight system.

Recommendation 2.5: Federal policy should require research ethics review that is commensurate with the nature and level of risk involved. Standards and procedures for review should distinguish between research that poses minimal risk and research that poses more than minimal risk. Minimal risk should be defined as the probability and magnitude of harms that are normally encountered in the daily lives of the general population (see Recommendation 4.2). In addition, the federal government should facilitate the creation of special, supplementary review bodies for research that involves novel or controversial ethical issues.

Recommendation 3.1: All institutions and sponsors engaged in research involving human participants should provide educational programs in research ethics to appropriate institutional officials, investigators, Institutional Review Board members, and Institutional Review Board staff. Among other issues, these programs should emphasize the obligations of institutions, sponsors, Institutional Review Boards, and investigators to protect the rights and welfare of participants. Colleges and universities should include research ethics in curricula related to research methods, and professional societies should include research ethics in their continuing education programs.

Recommendation 3.2: The federal government, in partnership with academic and professional societies, should enhance research ethics education related to protecting human research participants and stimulate the development of innovative educational programs. Professional societies should be consulted so that educational programs are designed to meet the needs of all who conduct and review research.

Recommendation 3.3: All investigators, Institutional Review Board members, and Institutional Review Board staff should be certified prior to conducting or reviewing research involving human participants. Certification requirements should be appropriate to their roles and to the area of research.

The federal government should encourage organizations, sponsors, and institutions to develop certification programs and mechanisms to evaluate their effectiveness. Federal policy should set standards for determining whether institutions and sponsors have an effective process of certification in place.

Recommendation 3.4: Sponsors, institutions, and independent Institutional Review Boards should be accredited in order to conduct or review research involving human participants. Accreditation should be premised upon demonstrated competency in core areas through accreditation programs that are approved by the federal government.

Recommendation 3.5: The process for assuring compliance with federal policy should be modified to reduce any unnecessary burden on institutions conducting research and to register institutions and Institutional Review Boards with the federal government. The assurance process should not be duplicative of accreditation programs for institutions (see Recommendation 3.4).

Recommendation 3.6: Institutions should develop internal mechanisms to ensure Institutional Review Board compliance and investigator compliance with regulations, guidance, and institutional procedures. Mechanisms should be put in place for reporting noncompliance to all relevant parties.

Recommendation 3.7: Federal policy should define institutional, Institutional Review Board, and investigator conflicts of interest, and guidance should be issued to ensure that the rights and welfare of research participants are protected.

Recommendation 3.8: Sponsors and institutions should develop policies and mechanisms to identify and manage all types of institutional, Institutional Review Board, and investigator conflicts of interest. In particular, all relevant conflicts of interest should be disclosed to participants. Policies also should describe specific types of prohibited relationships.

Recommendation 3.9: Federal policy should establish standards and criteria for the selection of Institutional Review Board members. The distribution of Institutional Review Board members with relevant expertise and experience should be commensurate with the types of research reviewed by the Institutional Review Board (see Recommendation 3.10).

Recommendation 3.10: Institutional Review Boards should include members who represent the perspectives of participants, members who are unaffiliated with the institution, and members whose primary concerns are in nonscientific areas. An individual can fulfill one, two, or all three of these categories. For the purposes of both overall membership and quorum determinations 1) these persons should collectively represent at least 25 percent of the Institutional Review Board membership and 2) members from all of these categories should be represented each time an Institutional Review Board meets (see Recommendation 3.9).

Recommendation 4.1: An analysis of the risks and potential benefits of study components should be applied to all types of covered research (see Recommendation 2.4). In general, each component of a study should be evaluated separately, and its risks should be both reasonable in themselves as well as justified by the potential benefits to society or the participants. Potential benefits from one component of a study should not be used to justify risks posed by a separate component of a study.

Recommendation 4.2: Federal policy should distinguish between research studies that pose minimal risk and those that pose more than minimal risk (see Recommendation 2.5). Minimal risk should be defined as the probability and magnitude of harms that are normally encountered in the daily lives of the general population. If a study that would normally be considered minimal risk for the general population nonetheless poses higher risk for any prospective participants, then the Institutional Review Board should approve the study only if it has determined that appropriate protections are in place for all prospective participants.

Recommendation 4.3: Federal policy should promote the inclusion of all segments of society in research. Guidance should be developed on how to identify and avoid situations that render some participants or groups vulnerable to harm or coercion. Sponsors and investigators should design research that incorporates appropriate safeguards to protect all prospective participants.

Recommendation 5.1: Federal policy should emphasize the process of informed consent rather than the form of its documentation and should ensure that competent participants have given their voluntary informed consent. Guidance should be issued about how to provide appropriate information to prospective research participants, how to promote prospective participants' comprehension of such information, and how to ensure that participants continue to make informed and voluntary decisions throughout their involvement in the research.

Recommendation 5.2: Federal policy should permit Institutional Review Boards in certain, limited situations (e.g., some studies using existing identifiable data or some observational studies) to waive informed consent requirements if all of the following criteria are met:

- all components of the study involve minimal risk or any component involving more than minimal risk must also offer the prospect of direct benefit to participants;
- the waiver is not otherwise prohibited by state, federal, or international law;
- there is an adequate plan to protect the confidentiality of the data;
- there is an adequate plan for contacting participants with information derived from the research, should the need arise; and

- in analyzing risks and potential benefits, the Institutional Review Board specifically determines that the benefits from the knowledge to be gained from the research study outweigh any dignitary harm associated with not seeking informed consent.

Recommendation 5.3: Federal policy should require investigators to document that they have obtained voluntary informed consent, but should be flexible with respect to the form of such documentation. Especially when individuals can easily refuse or discontinue participation, or when signed forms might threaten confidentiality, Institutional Review Boards should permit investigators to use other means of verifying that informed consent has been obtained.

Recommendation 5.4: Federal policy should be developed and mechanisms should be provided to enable investigators and institutions to reduce threats to privacy and breaches of confidentiality. The feasibility of additional mechanisms should be examined to strengthen confidentiality protections in research studies.

Recommendation 6.1: Federal policy should describe how sponsors, institutions, and investigators should monitor ongoing research.

Recommendation 6.2: Federal policy should describe clearly the requirements for continuing Institutional Review Board review of ongoing research. Continuing review should not be required for research studies involving minimal risk, research involving the use of existing data, or research that is in the data analysis phase when there is no additional contact with participants. When continuing review is not required, other mechanisms should be in place for ensuring compliance of investigators and for reporting protocol changes or unanticipated problems encountered in the research.

Recommendation 6.3: Federal policy should clarify when changes in research design or context require review and new approval by an Institutional Review Board.

Recommendation 6.4: The federal government should create a uniform system for reporting and evaluating adverse events occurring in research, especially in multi-site research. The reporting and evaluation responsibilities of investigators, sponsors, Institutional Review Boards, Data Safety Monitoring Boards, and federal agencies should be clear and efficient. The primary concern of the reporting system should be to protect current and prospective research participants.

Recommendation 6.5: For multi-site research, federal policy should permit central or lead Institutional Review Board review, provided that participants' rights and welfare are rigorously protected.

Recommendation 6.6: The federal government should study the issue of research-related injuries to determine if there is a need for a compensation program. If needed, the federal government should implement the recommendation of the President's Commission for the Study of Ethical Problems in Medicine and Biomedical and Behavioral Research (1982) to conduct a pilot study to evaluate possible program mechanisms.

Recommendation 7.1: The proposed oversight system should have adequate resources to ensure its effectiveness and ultimate success in protecting research participants and promoting research:

a) Funds should be appropriated to carry out the functions of the proposed federal oversight office as outlined in this report.
b) Federal appropriations for research programs should include a separate allocation for oversight activities related to the protection of human participants.
   - Institutions should be permitted to request funding for Institutional Review Boards and other oversight activities.
   - Federal agencies, other sponsors, and institutions should make additional funds available for oversight activities.

Recommendation 7.2: The federal government, in partnership with academic institutions and professional societies, should facilitate discussion about emerging human research protection issues and develop a research agenda that addresses issues related to research ethics.

CONCLUSIONS

This report proposes 30 recommendations for changing the oversight system at the national and local levels to ensure that all research participants receive the appropriate protections. The adoption of these recommendations, which are directed at all who are involved in the research enterprise, will not only lead to better protection for the participants of research, but will also serve to promote ethically sound research while reducing unnecessary bureaucratic burdens. Achieving these goals will, in turn, restore the respect of investigators for the system used to oversee research, support the public's trust in the research enterprise, and enhance public enthusiasm for all research involving human beings.

Source: Ethical and Policy Issues in Research Involving Human Participants. Bethesda, Md.: National Bioethics Advisory Commission, August 2001, pp. 9–18.

APPENDIX G

STEM CELLS AND THE FUTURE OF REGENERATIVE MEDICINE, 2002

Note: Listed below are the conclusions and recommendations from a report issued by the Committee on the Biological and Biomedical Applications of Stem Cell Research of the National Academy of Sciences.

- Experiments in mice and other animals are necessary, but not sufficient, for realizing the potential of stem cells to develop tissue-replacement therapies that will restore lost function in damaged organs. Because of the substantial biological differences between nonhuman animal and human development and between animal and human stem cells, studies with human stem cells are essential to make progress in the development of treatments for human disease, and this research should continue.
- There are important biological differences between adult and embryonic stem cells and among adult stem cells found in different types of tissue. The implications of these biological differences for therapeutic uses are not yet clear, and additional data are needed on all stem cell types. Adult stem cells from bone marrow have so far provided most of the examples of successful therapies for replacement of diseased or destroyed cells. Despite the enthusiasm generated by recent reports, the potential of adult stem cells to differentiate fully into other cell types (such as brain, nerve, pancreas cells) is still poorly understood and remains to be clarified. In contrast, studies of human embryonic stem cells have shown that they can develop into multiple tissue types and exhibit long-term self-renewal in culture, features that have not yet been demonstrated with many human adult stem cells. The application of stem cell research to

therapies for human disease will require much more knowledge about the biological properties of all types of stem cells. Although stem cell research is on the cutting edge of biological science today, it is still in its infancy. Studies of both embryonic and adult human stem cells will be required to most efficiently advance the scientific and therapeutic potential of regenerative medicine. Moreover, research on embryonic stem cells will be important to inform research on adult stem cells, and vice versa. Research on both adult and embryonic human stem cells should be pursued.

- Over time, all cell lines in tissue culture change, typically accumulating harmful genetic mutations. There is no reason to expect stem cell lines to behave differently. In addition, most existing stem cell lines have been cultured in the presence of non-human cells or serum that could lead to potential human health risks. Consequently, while there is much that can be learned using existing stem cell lines if they are made widely available for research, such concerns necessitate continued monitoring of these cells as well as the development of new stem cell lines in the future.
- High-quality, publicly funded research is the wellspring of medical breakthroughs. Although private, for-profit research plays a critical role in translating the fruits of basic research into medical advances that are broadly available to the public, stem cell research is far from the point of providing therapeutic products. Without public funding of basic research on stem cells, progress toward medical therapies is likely to be hindered. In addition, public funding offers greater opportunities for regulatory oversight and public scrutiny of stem cell research. Stem cell research that is publicly funded and conducted under established standards of open scientific exchange, peer review, and public oversight offers the most efficient and responsible means of fulfilling the promise of stem cells to meet the need for regenerative medical therapies.
- Conflicting ethical perspectives surround the use of embryonic stem cells in medical research, particularly where the moral and legal status of human embryos is concerned. The use of embryonic stem cells is not the first biomedical research activity to raise ethical and social issues among the public. Restrictions and guidelines for the conduct of controversial research have been developed to address such concerns in other instances. For example, when recombinant-DNA techniques raised questions and were subject to intense debate and public scrutiny, a national advisory body, the Recombinant DNA Advisory Committee, was established at the National Institutes of Health (NIH) to ensure

that the research met the highest scientific and ethical standards. If the federal government chooses to fund research on human embryonic stem cells, a similar national advisory group composed of exceptional researchers, ethicists, and other stakeholders should be established at NIH to oversee it. Such a group should ensure that proposals to work on human embryonic stem cells are scientifically justified and should scrutinize such proposals for compliance with federally mandated ethical guidelines.

- Regenerative medicine is likely to involve the implantation of new tissue in patients with damaged or diseased organs. A substantial obstacle to the success of transplantation of any cells, including stem cells and their derivatives, is the immune-mediated rejection of foreign tissue by the recipient's body. In current stem cell transplantation procedures with bone marrow and blood, success can hinge on obtaining a close match between donor and recipient tissues and on the use of immunosuppressive drugs, which often have severe and life-threatening side effects. To ensure that stem cell–based therapies can be broadly applicable for many conditions and individuals, new means to overcome the problem of tissue rejection must be found. Although ethically controversial, somatic cell nuclear transfer, a technique that produces a lineage of stem cells that are genetically identical to the donor, promises such an advantage. Other options for this purpose include genetic manipulation of the stem cells and the development of a very large bank of embryonic stem cell lines. In conjunction with research on stem cell biology and the development of stem cell therapies, research on approaches that prevent immune rejection of stem cells and stem cell–derived tissues should be actively pursued.

The committee is aware of and respectful of the wide array of social, political, legal, ethical, and economic issues that must be considered in policy-making in a democracy. And it is impressed by the commitment of all parties in this debate to life and health, regardless of the different conclusions they draw. The committee hopes that this report, by clarifying what is known about the scientific potential of stem cells and how that potential can best be realized, will be a useful contribution to the debate and to the enhancement of treatments for disabling human diseases and injuries. On August 9, 2001, when President Bush announced a new federal policy permitting limited use of human embryonic stem cells for research, this report was already in review. Because this report presents the committee's interpretation of the state of the science of stem cells independent of any specific policy, only minor modifications to refer to the new policy have been made in the report.

RECOMMENDATIONS

1. Studies with *human* stem cells are essential to make progress in the development of treatments for *human* disease, and this research should continue.
2. Although stem cell research is on the cutting edge of biological science today, it is still in its infancy. Studies of both embryonic and adult human stem cells will be required to most efficiently advance the scientific and therapeutic potential of regenerative medicine. Research on both adult and embryonic human stem cells should be pursued.
3. While there is much that can be learned using existing stem cell lines if they are made widely available for research, concerns about changing genetic and biological properties of these stem cell lines necessitate continued monitoring as well as the development of new stem cell lines in the future.
4. Human stem cell research that is publicly funded and conducted under established standards of open scientific exchange, peer review, and public oversight offers the most efficient and responsible means to fulfill the promise of stem cells to meet the need for regenerative medical therapies.
5. If the federal government chooses to fund human stem cell research, proposals to work on human embryonic stem cells should be required to justify the decision on scientific grounds and should be strictly scrutinized for compliance with existing and future federally mandated ethical guidelines.
6. A national advisory group composed of exceptional researchers, ethicists, and other stakeholders should be established at the National Institutes of Health (NIH) to oversee research on human embryonic stem cells. The group should include leading experts in the most current scientific knowledge relevant to stem cell research who can evaluate the technical merit of any proposed research on human embryonic stem cells. Other roles for the group could include evaluation of potential risks to research subjects and ensuring compliance with all legal requirements and ethical standards.
7. In conjunction with research on stem cell biology and the development of potential stem cell therapies, research on approaches that

prevent immune rejection of stem cells and stem cell–derived tissues should be actively pursued. These scientific efforts include the use of a number of techniques to manipulate the genetic makeup of stem cells, including somatic cell nuclear transfer.

Source: Committee on the Biological and Biomedical Applications of Stem Cell Research, Board on Life Sciences National Research Council. *Stem Cells and the Future of Regenerative Medicine.* Washington, D.C.: National Academy Press, 2002, pp. 2–5.

APPENDIX H

GUIDELINES FOR HUMAN EMBRYONIC STEM CELL RESEARCH, 2005

Note: In 2005, the Committee on Guidelines for Human Embryonic Stem Cell Research of the Board on the Life Sciences of the National Research Council and the Health Sciences Policy Board of the Institute of Medicine issued its report on policies for human embryonic stem cell (hES) research. A summary of it recommendations follows.

APPENDIX A: COMPILATION OF RECOMMENDATIONS

RECOMMENDATIONS FROM CHAPTER 3

Recommendation 1

To provide local oversight of all issues related to derivation and research use of hES cell lines and to facilitate education of investigators involved in hES cell research, each institution should establish an Embryonic Stem Cell Research Oversight (ESCRO) committee. The committee should include representatives of the public and persons with expertise in developmental biology, stem cell research, molecular biology, assisted reproduction, and ethical and legal issues in hES cell research. The ESCRO committee would not substitute for an Institutional Review Board but rather would provide an additional level of review and scrutiny warranted by the complex issues raised by hES cell research. The committee would also serve to review basic hES cell research using preexisting anonymous cell lines that does not require consideration by an Institutional Review Board.

Recommendation 2

Through its Embryonic Stem Cell Research Oversight (ESCRO) committee, each research institution should ensure that the provenance of hES cells is documented. Documentation should include evidence that the procurement process was approved by an Institutional Review Board to ensure adherence to the basic ethical and legal principles of informed consent and protection of confidentiality.

Recommendation 3

Embryonic Stem Cell Research Oversight (ESCRO) committees or their equivalents should divide research proposals into three categories in setting limits on research and determining the requisite level of oversight:

(a) Research that is permissible after notification of the research institution's ESCRO committee and completion of the reviews mandated by current requirements. Purely in vitro hES cell research with preexisting coded or anonymous hES cell lines in general is permissible provided that notice of the research, documentation of the provenance of the cell lines, and evidence of compliance with any required Institutional Review Board, Institutional Animal Care and Use Committee, Institutional Biosafety Committee, or other mandated reviews is provided to the ESCRO committee or other body designated by the investigator's institution.

(b) Research that is permissible only after additional review and approval by an ESCRO committee or other equivalent body designated by the investigator's institution. The ESCRO committee should evaluate all requests for permission to attempt derivation of new hES cell lines from donated blastocysts, from in vitro fertilized oocytes, or by nuclear transfer. The scientific rationale for the need to generate new hES cell lines, by whatever means, must be clearly presented, and the basis for the numbers of blastocysts and oocytes needed should be justified. Such requests should be accompanied by evidence of Institutional Review Board approval of the procurement process. All research involving the introduction of hES cells into nonhuman animals at any stage of embryonic, fetal, or postnatal development should be reviewed by the ESCRO committee. Particular attention should be paid to the probable pattern and effects of differentiation and integration of the human cells into the nonhuman animal tissues. Research in which personally identifiable information about the donors of the blastocysts, gametes, or somatic cells from which the hES cells were derived is readily ascertainable by the investigator also requires ESCRO committee review and approval.

(c) Research that should not be permitted at this time:
   i) Research involving in vitro culture of any intact human embryo, regardless of derivation method, for longer than 14 days or until formation of the primitive streak begins, whichever occurs first.
   ii) Research in which hES cells are introduced into nonhuman primate blastocysts or in which any ES cells are introduced into human blastocysts. in addition:
   iii) No animal into which hES cells have been introduced at any stage of development should be allowed to breed.

Recommendation 4

Through its Embryonic Stem Cell Research Oversight (ESCRO) committee, each research institution should establish and maintain a registry of investigators conducting hES cell research and record descriptive information about the types of research being performed and the hES cells in use.

Recommendation 5

If a U.S.-based investigator collaborates with an investigator in another country, the Embryonic Stem Cell Research Oversight (ESCRO) committee may determine that the procedures prescribed by the foreign institution afford protections equivalent with these guidelines and may approve the substitution of some or all of the foreign procedures for its own.

Recommendation 6

A national body should be established to assess periodically the adequacy of the guidelines proposed in this document and to provide a forum for a continuing discussion of issues involved in hES cell research.

Recommendation 7

The hES cell research community should ensure that there is sufficient genetic diversity among cell lines to allow for potential translation into health care services for all groups in our society.

RECOMMENDATIONS FROM CHAPTER 4

Recommendation 8

Regardless of the source of funding and the applicability of federal regulations, an Institutional Review Board or its equivalent should review the procurement of gametes, blastocysts, or somatic cells for the purpose of generating new hES cell lines, including the procurement of blastocysts in excess of clinical need from in vitro fertilization clinics, blastocysts made through in vitro fertilization specifically for research purposes, and oocytes, sperm, and somatic cells donated for development of hES cell lines derived through nuclear transfer.

Recommendation 9

Institutional Review Boards may not waive the requirement for obtaining informed consent from any person whose somatic cells, gametes, or blastocysts are used in hES research.

Recommendation 10

Investigators, institutions, Institutional Review Boards, and privacy boards should ensure that authorizations are received from donors, as appropriate and required by federal human subjects protections and the Health Insurance Portability and Accountability Act for the confidential transmission of personal health information to repositories or to investigators who are using hES cell lines derived from donated materials.

Recommendation 11

Investigators and institutions involved in hES cell research should conduct the research in accordance with all applicable laws and guidelines pertaining to recombinant DNA research and animal care. Institutions should consider adopting Good Laboratory Practice standards for some or all of their basic hES cell research.

Recommendation 12

hES cell research leading to potential clinical application must be in compliance with all applicable Food and Drug Administration (FDA) regulations. If FDA requires that a link to the donor source be maintained, investigators and institutions must ensure that the confidentiality of the donor is protected, that the donor understands that a link will be maintained, and that, where applicable, federal human subjects protections and Health Insurance Portability and Accountability Act or other privacy protections are followed.

Recommendations from Chapter 5

Recommendation 13

When donor gametes have been used in the in vitro fertilization process, resulting blastocysts may not be used for research without consent of all gamete donors.

Appendix H

Recommendation 14

To facilitate autonomous choice, decisions related to the production of embryos for infertility treatment should be free of the influence of investigators who propose to derive or use hES cells in research. Whenever it is practicable, the attending physician responsible for the infertility treatment and the investigator deriving or proposing to use hES cells should not be the same person.

Recommendation 15

No cash or in kind payments may be provided for donating blastocysts in excess of clinical need for research purposes.

Recommendation 16

Women who undergo hormonal induction to generate oocytes specifically for research purposes (such as for nuclear transfer) should be reimbursed only for direct expenses incurred as a result of the procedure, as determined by an Institutional Review Board. No cash or in kind payments should be provided for donating oocytes for research purposes. Similarly, no payments should be made for donations of sperm for research purposes or of somatic cells for use in nuclear transfer.

Recommendation 17

Consent for blastocyst donation should be obtained from each donor at the time of donation. Even people who have given prior indication of their intent to donate to research any blastocysts that remain after clinical care should nonetheless give informed consent at the time of donation. Donors should be informed that they retain the right to withdraw consent until the blastocysts are actually used in cell line derivation.

Recommendation 18

In the context of donation of gametes or blastocysts for hES cell research, the informed consent process, should, at a minimum, provide the following information:

a. A statement that the blastocysts or gametes will be used to derive hES cells for research that may include research on human transplantation.
b. A statement that the donation is made without any restriction or direction regarding who may be the recipient of transplants of the cells derived, except in the case of autologous donation.

c. A statement as to whether the identities of the donors will be readily ascertainable to those who derive or work with the resulting hES cell lines.
d. If the identities of the donors are retained (even if coded), a statement as to whether donors wish to be contacted in the future to receive information obtained through studies of the cell lines.
e. An assurance that participants in research projects will follow applicable and appropriate best practices for donation, procurement, culture, and storage of cells and tissues to ensure, in particular, the traceability of stem cells. (Traceable information, however, must be secured to ensure confidentiality.)
f. A statement that derived hES cells and/or cell lines might be kept for many years.
g. A statement that the hES cells and/or cell lines might be used in research involving genetic manipulation of the cells or the mixing of human and nonhuman cells in animal models.
h. Disclosure of the possibility that the results of study of the hES cells may have commercial potential and a statement that the donor will not receive financial or any other benefits from any future commercial development.
i. A statement that the research is not intended to provide direct medical benefit to the donor(s) except in the case of autologous donation.
j. A statement that embryos will be destroyed in the process of deriving hES cells.
k. A statement that neither consenting nor refusing to donate embryos for research will affect the quality of any future care provided to potential donors.
l. A statement of the risks involved to the donor.

Recommendation 19

Consenting or refusing to donate gametes or embryos for research should not affect or alter in any way the quality of care provided to prospective donors. That is, clinical staff must provide appropriate care to patients without prejudice regarding their decisions about disposition of their embryos.

Recommendation 20

Clinical personnel who have a conscientious objection to hES cell research should not be required to participate in providing donor information or securing donor consent for research use of gametes or blastocysts. That privilege should not extend to the care of a donor or recipient.

Appendix H

Recommendation 21

Researchers may not ask members of the infertility treatment team to generate more oocytes than necessary for the optimal chance of reproductive success. An infertility clinic or other third party responsible for obtaining consent or collecting materials should not be able to pay for or be paid for the material obtained (except for specifically defined cost-based reimbursements and payments for professional services).

Recommendation 22

Institutions that are banking or plan to bank hES cell lines should establish uniform guidelines to ensure that donors of material give informed consent through a process approved by an Institutional Review Board, and that meticulous records are maintained about all aspects of cell culture. Uniform tracking systems and common guidelines for distribution of cells should be established.

Recommendation 23

Any facility engaged in obtaining and storing hES cell lines should consider the following standards:

a. Creation of a committee for policy and oversight purposes and creation of clear and standardized protocols for banking and withdrawals.
b. Documentation requirements for investigators and sites that deposit cell lines, including
    i. A copy of the donor consent form.
    ii. Proof of Institutional Review Board approval of the procurement process.
    iii. Available medical information on the donors, including results of infectious-disease screening.
    iv. Available clinical, observational, or diagnostic information about the donor(s).
    v. Critical information about culture conditions (such as media, cell passage, and safety information).
    vi. Available cell line characterization (such as karyotype and genetic markers). A repository has the right of refusal if prior culture conditions or other items do not meet its standards.
c. A secure system for protecting the privacy of donors when materials retain codes or identifiable information, including but not limited to
    i. A schema for maintaining confidentiality (such as a coding system).

ii. A system for a secure audit trail from primary cell lines to those submitted to the repository.
iii. A policy governing whether and how to deliver clinically significant information back to donors.

d. The following standard practices:
i. Assignment of a unique identifier to each sample.
ii. A process for characterizing cell lines.
iii. A process for expanding, maintaining, and storing cell lines.
iv. A system for quality assurance and control.
v. A Website that contains scientific descriptions and data related to the cell lines available.
vi. A procedure for reviewing applications for cell lines.
vii. A process for tracking disbursed cell lines and recording their status when shipped (such as number of passages).
viii. A system for auditing compliance.
ix. A schedule of charges.
x. A statement of intellectual property policies.
xi. When appropriate, creation of a clear Material Transfer Agreement or user agreement.
xii. A liability statement.
xiii. A system for disposal of material.

e. Clear criteria for distribution of cell lines, including but not limited to evidence of approval of the research by an embryonic stem cell research oversight committee or equivalent body at the recipient institution.

Source: Committee on Guidelines for Human Embryonic Stem Cell Research, Board on Life Sciences, National Research Council and Health Sciences Policy Board, Institute of Medicine. *Guidelines for Human Embryonic Stem Cell Research.* Washington, D.C. National Academies Press, 2005, pp. 107–112.

APPENDIX I

LEGALIZATION OF STEM CELL RESEARCH IN NEW JERSEY, JANUARY 4, 2005

CHAPTER 203

AN ACT concerning human stem cell research and supplementing Title 26 of the Revised Statutes and Title 2C of the New Jersey Statutes.

BE IT ENACTED by the Senate and General Assembly of the State of New Jersey

C.26:2Z-1 Findings, declarations relative to human stem cell research.

1. The Legislature finds and declares that:

a. An estimated 128 million Americans suffer from the crippling economic and psychological burden of chronic, degenerative and acute diseases, including Alzheimer's disease, cancer, diabetes and Parkinson's disease;

b. The costs of treating, and lost productivity from, chronic, degenerative and acute diseases in the United States constitutes hundreds of billions of dollars annually. Estimates of the economic costs of these diseases does not account for the extreme human loss and suffering associated with these conditions;

c. Human stem cell research offers immense promise for developing new medical therapies for these debilitating diseases and a critical means to explore fundamental questions of biology. Stem cell research could lead to unprecedented treatments and potential cures for Alzheimer's disease, cancer, diabetes, Parkinson's disease and other diseases;

d. The United States has historically been a haven for open scientific inquiry and technological innovation; and this environment, combined with the commitment of public and private resources, has made this nation the preeminent world leader in biomedicine and biotechnology;

e. The biomedical industry is a critical and growing component of New Jersey's economy, and would be significantly diminished by limitations imposed on stem cell research;

f. Open scientific inquiry and publicly funded research will be essential to realizing the promise of stem cell research and maintaining this State's leadership in biomedicine and biotechnology. Publicly funded stem cell research, conducted under established standards of open scientific exchange, peer review and public oversight, offers the most efficient and responsible means of fulfilling the promise of stem cells to provide regenerative medical therapies;

g. Stem cell research, including the use of embryonic stem cells for medical research, raises significant ethical and public policy concerns; and, although not unique, the ethical and policy concerns associated with stem cell research must be carefully considered; and

h. The public policy of this State governing stem cell research must: balance ethical and medical considerations, based upon both an understanding of the science associated with stem cell research and a thorough consideration of the ethical concerns regarding this research; and be carefully crafted to ensure that researchers have the tools necessary to fulfill the promise of this research.

C.26:2Z-2 Public policy relative to derivation, use of certain cells of humans.

2. a. It is the public policy of this State that research involving the derivation and use of human embryonic stem cells, human embryonic germ cells and human adult stem cells, including somatic cell nuclear transplantation, shall:

(1) be permitted in this State;

(2) be conducted with full consideration for the ethical and medical implications of this research; and

(3) be reviewed, in each case, by an institutional review board operating in accordance with applicable federal regulations.

b. (1) A physician or other health care provider who is treating a patient for infertility shall provide the patient with timely, relevant and appropriate information sufficient to allow that person to make an informed and voluntary choice regarding the disposition of any human embryos remaining following the infertility treatment.

(2) A person to whom information is provided pursuant to paragraph (1) of this subsection shall be presented with the option of storing any unused embryos, donating them to another person, donating the remaining embryos for research purposes, or other means of disposition.

(3) A person who elects to donate, for research purposes, any embryos remaining after receiving infertility treatment shall provide written consent to that donation.

c. (1) A person shall not knowingly, for valuable consideration, purchase or sell, or otherwise transfer or obtain, or promote the sale or transfer of, embryonic or cadaveric fetal tissue for research purposes pursuant to this act; however, embryonic or cadaveric fetal tissue may be donated for research purposes in accordance with the provisions of subsection b. of this section or other applicable State or federal law.

For the purposes of this subsection, "valuable consideration" means financial gain or advantage, but shall not include reasonable payment for the removal, processing, disposal, preservation, quality control, storage, transplantation, or implantation of embryonic or cadaveric fetal tissue.

(2) A person or entity who violates the provisions of this subsection shall be guilty of a crime of the third degree and, notwithstanding the provisions of subsection b. of N.J.S.2C:43–3, shall be subject to a fine of up to $50,000 for each violation.

C.2C11A-1 Cloning of human being, first degree crime; definition.

3. A person who knowingly engages or assists, directly or indirectly, in the cloning of a human being is guilty of a crime of the first degree.

As used in this section, "cloning of a human being" means the replication of a human individual by cultivating a cell with genetic material through the egg, embryo, fetal and newborn stages into a new human individual.

4. This act shall take effect immediately.

Approved January 2, 2004.

Source: New Jersey Statutes. Available online. URL: http//www.state.nj.us/scitech/stem%20cell%20final%20bill.pdf. Downloaded on July 22, 2005.

APPENDIX J

STEM CELL ENHANCEMENT ACT, FEBRUARY 15, 2005

109th CONGRESS

H. R. 810

1st SESSION

To amend the Public Health Service Act to provide for human embryonic stem cell research.

IN THE HOUSE OF REPRESENTATIVES

FEBRUARY 15, 2005

Mr. CASTLE (for himself, Ms. DEGETTE [co-sponsor], [and 153 named other sponsors])

A BILL

To amend the Public Health Service Act to provide for human embryonic stem cell research.

Be it enacted by the Senate and House of Representatives of the United States of America in Congress assembled,

SECTION 1. SHORT TITLE.

This Act may be cited as the "Stem Cell Research Enhancement Act of 2005".

SEC. 2. HUMAN EMBRYONIC STEM CELL RESEARCH.

Part H of title IV of the Public Health Service Act (42 U.S.C. 289 et seq.) is amended by inserting after section 498C the following:

"SEC. 498D. HUMAN EMBRYONIC STEM CELL RESEARCH.

"(a) IN GENERAL.—Notwithstanding any other provision of law (including any regulation or guidance), the Secretary shall conduct and support research that utilizes human embryonic stem cells in accordance with this section (regardless of the date on which the stem cells were derived from a human embryo).

"(b) ETHICAL REQUIREMENTS.—1 .—Human embryonic stem cells shall be eligible for use in any research conducted or supported by the Secretary if the cells meet each of the following:

"(1) The stem cells were derived from human embryos that have been donated from in vitro fertilization clinics, were created for the purposes of fertility treatment, and were in excess of the clinical need of the individuals seeking such treatment.

"(2) Prior to the consideration of embryo donation and through consultation with the individuals seeking fertility treatment, it was determined that the embryos would never be implanted in a woman and would otherwise be discarded.

"(3) The individuals seeking fertility treatment donated the embryos with written informed consent and without receiving any financial or other inducements to make the donation.

"(c) GUIDELINES.—Not later than 60 days after the date of the enactment of this section, the Secretary, in consultation with the Director of NIH, shall issue final guidelines to carry out this section.

"(d) REPORTING REQUIREMENTS.—The Secretary shall annually prepare and submit to the appropriate committees of the Congress a report describing the activities carried out under this section during the preceding fiscal year, and including a description of whether and to what extent research under subsection (a) has been conducted in accordance with this section.".

Source: "Stem Cell Enhancement Act of 2005." Thomas, Library of Congress. Available online. URL: http://thomas.loc.gov/cgi-bin/query/D?c109:4:./temp/~c109GoQ7pD::. Downloaded on July 21, 2005.

APPENDIX K

HUMAN CLONING PROHIBITION ACT OF 2005, MARCH 17, 2005

S. 658

A BILL

To amend the Public Health Service Act to prohibit human cloning.

Be it enacted by the Senate and House of Representatives of the United States of America in Congress assembled,

SECTION 1. SHORT TITLE.

This Act may be cited as the 'Human Cloning Prohibition Act of 2005'.

SEC. 2. PROHIBITION ON HUMAN CLONING.

Part H of title IV of the Public Health Service Act (42 U.S.C. 289 et seq.) is amended by adding at the end the following:

'SEC. 498D. PROHIBITION ON HUMAN CLONING.

'(a) Definitions—In this section:

'(1) HUMAN CLONING—The term 'human cloning' means human asexual reproduction, accomplished by introducing nuclear material from one or more human somatic cells into a fertilized or unfertilized oocyte whose nuclear material has been removed or inactivated so as to produce a living organism (at any stage of development) that is genetically virtually identical to an existing or previously existing human organism.

'(2) ASEXUAL REPRODUCTION—The term 'asexual reproduction' means reproduction not initiated by the union of oocyte and sperm.

'(3) SOMATIC CELL—The term 'somatic cell' means a diploid cell (having a complete set of chromosomes) obtained or derived from a living or deceased human body at any stage of development.

'(b) Prohibition—It shall be unlawful for any person or entity, public or private, in or affecting interstate commerce, knowingly —

'(1) to perform or attempt to perform human cloning;

'(2) to participate in an attempt to perform human cloning; or

'(3) to ship or receive for any purpose an embryo produced by human cloning or any product derived from such embryo.

'(c) Importation—It shall be unlawful for any person or entity, public or private, knowingly to import for any purpose an embryo produced by human cloning.

'(d) Penalties —

'(1) CRIMINAL PENALTY—Any person or entity that violates this section shall be fined or imprisoned for not more than 10 years, or both.

'(2) CIVIL PENALTY—Any person or entity that violates any provision of this section shall be subject to, in the case of a violation that involves the derivation of a pecuniary gain, a civil penalty of not less than $1,000,000 and not more than an amount equal to the amount of the gross gain multiplied by 2, if that amount is greater than $1,000,000.

'(e) Scientific Research—Nothing in this section restricts areas of scientific research not specifically prohibited by this section, including research in the use of nuclear transfer or other cloning techniques to produce molecules, DNA, cells other than human embryos, tissues, organs, plants, or animals other than humans.'.

SEC. 3. STUDY BY GOVERNMENT ACCOUNTABILITY OFFICE.

(a) In General—The Government Accountability Office shall conduct a study to assess the need for amendment of the prohibition on human cloning, as defined in section 498D(a) of the Public Health Service Act, as added by section 2, which study should include —

(1) a discussion of new developments in medical technology concerning human cloning and somatic cell nuclear transfer, the need (if any) for somatic cell nuclear transfer to produce medical advances, current public attitudes and prevailing ethical views concerning the use of somatic cell nuclear transfer, and potential legal implications of research in somatic cell nuclear transfer; and

(2) a review of any technological developments that may require that technical changes be made to section 498D of the Public Health Service Act.

(b) Report—The Government Accountability Office shall transmit to Congress, not later than 4 years after the date of enactment of this Act, a report containing the findings and conclusions of its study, together with recommendations for any legislation or administrative actions which it considers appropriate.

Source: "Human Cloning Prohibition Act of 2005." Thomas, Library of Congress. Available online. URL: http://thomas.loc.gov/cgi-bin/query/z?c109:S.658:. Downloaded on July 21, 2005.

APPENDIX L

FEDERAL LAWS CONCERNING EXPERIMENTATION ON EMBRYOS AND FETUSES, AS OF 2005

CODE OF FEDERAL REGULATIONS

TITLE 45—PUBLIC WELFARE AND HUMAN SERVICES
PART 46—PROTECTION OF HUMAN SUBJECTS—Table of Contents

Subpart B—Additional Protections Pertaining to Research, Development, and Related Activities Involving Fetuses, Pregnant Women, and Human In Vitro Fertilization

Sec. 46.208 Activities directed toward fetuses in utero as subjects.

(a) No fetus in utero may be involved as a subject in any activity covered by this subpart unless: (1) The purpose of the activity is to meet the health needs of the particular fetus and the fetus will be placed at risk only to the minimum extent necessary to meet such needs, or (2) the risk to the fetus imposed by the research is minimal and the purpose of the activity is the development of important biomedical knowledge which cannot be obtained by other means.

(b) An activity permitted under paragraph (a) of this section may be conducted only if the mother and father are legally competent and have given their informed consent, except that the father's consent need not be secured if: (1) His identity or whereabouts cannot reasonably be ascertained, (2) he is not reasonably available, or (3) the pregnancy resulted from rape.

Source: Code of Federal Regulations. Available online. URL: http://www.hhs.gov/ohrp/humansubjects/guidance/45cfr46.htm#subpartb. Downloaded on July 22, 2005.

UNITED STATES CODE

TITLE 42
CHAPTER 6A
SUBCHAPTER III
Part H
Section 289g-1. Research on transplantation of fetal tissue

(a) Establishment of program

(1) In general

The Secretary may conduct or support research on the transplantation of human fetal tissue for therapeutic purposes.

(2) Source of tissue

Human fetal tissue may be used in research carried out under paragraph (1) regardless of whether the tissue is obtained pursuant to a spontaneous or induced abortion or pursuant to a stillbirth.

(b) Informed consent of donor

(1) In general

In research carried out under subsection (a) of this section, human fetal tissue may be used only if the woman providing the tissue makes a statement, made in writing and signed by the woman, declaring that —

(A) the woman donates the fetal tissue for use in research described in subsection (a) of this section;

(B) the donation is made without any restriction regarding the identity of individuals who may be the recipients of transplantations of the tissue; and

(C) the woman has not been informed of the identity of any such individuals.

(2) Additional statement

In research carried out under subsection (a) of this section, human fetal tissue may be used only if the attending physician with respect to obtaining the tissue from the woman involved makes a statement, made in writing and signed by the physician, declaring that —

(A) in the case of tissue obtained pursuant to an induced abortion —

(i) the consent of the woman for the abortion was obtained prior to requesting or obtaining consent for a donation of the tissue for use in such research;

(ii) no alteration of the timing, method, or procedures used to terminate the pregnancy was made solely for the purposes of obtaining the tissue; and
(iii) the abortion was performed in accordance with applicable State law;

(B) the tissue has been donated by the woman in accordance with paragraph (1); and
(C) full disclosure has been provided to the woman with regard to —

(i) such physician's interest, if any, in the research to be conducted with the tissue; and
(ii) any known medical risks to the woman or risks to her privacy that might be associated with the donation of the tissue and that are in addition to risks of such type that are associated with the woman's medical care.

(c) Informed consent of researcher and donee
in research carried out under subsection (a) of this section, human fetal tissue may be used only if the individual with the principal responsibility for conducting the research involved makes a statement, made in writing and signed by the individual, declaring that the individual —
(1) is aware that —

(A) the tissue is human fetal tissue;
(B) the tissue may have been obtained pursuant to a spontaneous or induced abortion or pursuant to a stillbirth; and
(C) the tissue was donated for research purposes;

(2) has provided such information to other individuals with responsibilities regarding the research;
(3) will require, prior to obtaining the consent of an individual to be a recipient of a transplantation of the tissue, written acknowledgment of receipt of such information by such recipient; and
(4) has had no part in any decisions as to the timing, method, or procedures used to terminate the pregnancy made solely for the purposes of the research.
(d) Availability of statements for audit
(1) In general
In research carried out under subsection (a) of this section, human fetal tissue may be used only if the head of the agency or other entity conducting the research involved certifies to the Secretary that the statements required under subsections

(b)(2) and (c) of this section will be available for audit by the Secretary.

(2) Confidentiality of audit

Any audit conducted by the Secretary pursuant to paragraph (1) shall be conducted in a confidential manner to protect the privacy rights of the individuals and entities involved in such research, including such individuals and entities involved in the donation, transfer, receipt, or transplantation of human fetal tissue. With respect to any material or information obtained pursuant to such audit, the Secretary shall —

(A) use such material or information only for the purposes of verifying compliance with the requirements of this section;

(B) not disclose or publish such material or information, except where required by Federal law, in which case such material or information shall be coded in a manner such that the identities of such individuals and entities are protected; and

(C) not maintain such material or information after completion of such audit, except where necessary for the purposes of such audit.

(e) Applicability of State and local law

(1) Research conducted by recipients of assistance

The Secretary may not provide support for research under subsection (a) of this section unless the applicant for the financial assistance involved agrees to conduct the research in accordance with applicable State law.

(2) Research conducted by Secretary

The Secretary may conduct research under subsection (a) of this section only in accordance with applicable State and local law.

(f) Report

The Secretary shall annually submit to the Committee on Energy and Commerce of the House of Representatives, and to the Committee on Labor and Human Resources of the Senate, a report describing the activities carried out under this section during the preceding fiscal year, including a description of whether and to what extent research under subsection (a) of this section has been conducted in accordance with this section.

(g) "Human fetal tissue" defined

For purposes of this section, the term "human fetal tissue" means tissue or cells obtained from a dead human embryo or fetus after a spontaneous or induced abortion, or after a stillbirth.

Source: US Code Collection. Available online. URL: http://www4.law.cornell.edu/uscode/html/uscode42/usc_sec_42_00000289—g001-.html. Downloaded on July 22, 2005.

APPENDIX M

California Law on Stem Cell Research, as of 2005

CALIFORNIA CODES
HEALTH AND SAFETY CODE
SECTIONS 125300–125320

125300. The policy of the State of California shall be as follows:

(a) That research involving the derivation and use of human embryonic stem cells, human embryonic germ cells, and human adult stem cells from any source, including somatic cell nuclear transplantation, shall be permitted and that full consideration of the ethical and medical implications of this research be given.

(b) That research involving the derivation and use of human embryonic stem cells, human embryonic germ cells, and human adult stem cells, including somatic cell nuclear transplantation, shall be reviewed by an approved institutional review board.

125305. (a) The department shall establish and maintain an anonymous registry of embryos that are available for research. The purpose of this registry is to provide researchers with access to embryos that are available for research purposes.

(b) The department may contract with the University of California, private organizations, or public entities to establish and administer the registry.

(c) This section shall be implemented only to the extent that funds for the purpose of establishing and administering the registry are received by the department from private or other nonstate sources.

125315. (a) A physician and surgeon or other health care provider delivering fertility treatment shall provide his or her patient with timely, relevant, and appropriate information to allow the individual to make an informed and voluntary choice regarding the disposition of any human embryos remaining following the fertility treatment. The failure to provide to a patient this information constitutes unprofessional conduct within the meaning of Chapter 5 (commencing with Section 2000) of Division 2 of the Business and Professions Code.

(b) Any individual to whom information is provided pursuant to subdivision (a) shall be presented with the option of storing any unused embryos, donating them to another individual, discarding the embryos, or donating the remaining embryos for research. When providing fertility treatment, a physician and surgeon or other health care provider shall provide a form to the male and female partner, or the individual without a partner, as applicable, that sets forth advanced written directives regarding the disposition of embryos. This form shall indicate the time limit on storage of the embryos at the clinic or storage facility and shall provide, at a minimum, the following choices for disposition of the embryos based on the following circumstances:

(1) In the event of the death of either the male or female partner, the embryos shall be disposed of by one of the following actions:

(A) Made available to the living partner.
(B) Donation for research purposes.
(C) Thawed with no further action taken.
(D) Donation to another couple or individual.
(E) Other disposition that is clearly stated.

(2) In the event of the death of both partners or the death of a patient without a partner, the embryos shall be disposed of by one of the following actions:

(A) Donation for research purposes.
(B) Thawed with no further action taken.
(C) Donation to another couple or individual.
(D) Other disposition that is clearly stated.

(3) In the event of separation or divorce of the partners, the embryos shall be disposed of by one of the following actions:

(A) Made available to the female partner.
(B) Made available to the male partner.
(C) Donation for research purposes.
(D) Thawed with no further action taken.

(E) Donation to another couple or individual.
(F) Other disposition that is clearly stated.

(4) In the event of the partners' decision or a patient's decision who is without a partner, to abandon the embryos by request or a failure to pay storage fees, the embryos shall be disposed of by one of the following actions:

(A) Donation for research purposes.
(B) Thawed with no further action taken.
(C) Donation to another couple or individual.
(D) Other disposition that is clearly stated.

(c) A physician and surgeon or other health care provider delivering fertility treatment shall obtain written consent from any individual who elects to donate embryos remaining after fertility treatments for research. For any individual considering donating the embryos for research, to obtain informed consent, the health care provider shall convey all of the following to the individual:

(1) A statement that the early human embryos will be used to derive human pluripotent stem cells for research and that the cells may be used, at some future time, for human transplantation research.
(2) A statement that all identifiers associated with the embryos will be removed prior to the derivation of human pluripotent stem cells.
(3) A statement that donors will not receive any information about subsequent testing on the embryo or the derived human pluripotent cells.
(4) A statement that derived cells or cell lines, with all identifiers removed, may be kept for many years.
(5) Disclosure of the possibility that the donated material may have commercial potential, and a statement that the donor will not receive financial or any other benefits from any future commercial development.
(6) A statement that the human pluripotent stem cell research is not intended to provide direct medical benefit to the donor.
(7) A statement that early human embryos donated will not be transferred to a woman's uterus, will not survive the human pluripotent stem cell derivation process, and will be handled respectfully, as is appropriate for all human tissue used in research.

125320 (a) A person may not knowingly, for valuable consideration, purchase or sell embryonic or cadaveric fetal tissue for research purposes pursuant to this chapter.

(b) For purposes of this section, "valuable consideration" does not include reasonable payment for the removal, processing, disposal, preservation, quality control, storage, transplantation, or implantation of a part.

(c) Embryonic or cadaveric fetal tissue may be donated for research purposes pursuant to this chapter.

Source: California Codes. Available online. URL: http://www.leginfo.ca.gov/cgi-bin/waisgate?WAISdocID=0499597823+0+0+0&WAISaction=retrieve. Downloaded on July 22, 2005.

INDEX

Locators in **boldface** indicate main topics. Locators followed by *c* indicate chronology entries. Locators followed by *b* indicate biographical entries. Locators followed by *g* indicate glossary entries. Locators followed by *t* indicate diagrams.

Index

Index

Index

Index

Index

Index

Index